Philosophy and Probability

Philosophy and Probability

Timothy Childers

OXFORD
UNIVERSITY PRESS

OXFORD
UNIVERSITY PRESS

Great Clarendon Street, Oxford, OX2 6DP,
United Kingdom

Oxford University Press is a department of the University of Oxford.
It furthers the University's objective of excellence in research, scholarship,
and education by publishing worldwide. Oxford is a registered trade mark of
Oxford University Press in the UK and in certain other countries

Impression: 1

British Library Cataloguing in Publication Data

Data available

ISBN 978-0-19-966182-4 (Hbk)
ISBN 978-0-19-966183-1 (Pbk)

Printed and bound by CPI Group (UK) Ltd, Croydon, CR0 4YY

For Dagmar, Laura, and Lukas

Contents

Detailed Contents

Preface

Probabilities permeate our lives: they show up in weather reports, science, popular reports of science, predictions about election results, chances for surviving diseases, prices on futures markets, and, of course, in casinos. Probability plays such an important role in modern life that it is no surprise that philosophers are interested in them.

This interest might seem a bit odd: probability is a part of mathematics—what do philosophers have to bring to a discussion about mathematics? But this is to misunderstand the subject. The probability calculus is a bit of mathematics, but its *interpretation* is another matter. Philosophy enquires into the meaning of probabilities, or, perhaps more accurately, into the meaning of statements about probability. And the probability calculus holds an interesting position: the calculus itself is well established, but none of its interpretations are.

Some examples: What does it mean that smoking raises your chance of getting cancer? What does it mean that the candidate for elected office has increased her chances of getting elected by having spent heavily in the campaign on advertisements? Are these two chances the same kind of chance? These are by no means trivial questions: indeed, after years of intense philosophical work, there is no general agreement. There are several schools of thought: in the next paragraphs we will introduce the most well known.

One school of thought, sometimes awkwardly called frequentism for the *relative frequency* interpretation, is that there is a relationship between smoking and cancer, and that this relationship can be described using probabilities. But, as to *your* chance of getting cancer, probability is silent, for probability does not deal in individuals, but groups only. Still, we can use probabilities to set the cost of your (and everyone else's) insurance. This school also holds that probability does not apply to a particular candidate's spending money in a particular campaign, and, in fact, we might not be able to determine any link between filthy lucre and high office.

A second school of thought, occupied mostly by advocates of the *propensity* interpretation, is that there *is* a chance, an objective chance, of your getting cancer. But as to whether we will ever know what the chance is, and whether that chance is the same for those similar to you is the subject of great controversy. The same goes for our big-spending candidate: it's objectively true that spending will, or will not, increase his or her chances, according to this school. But whether this holds for all candidates, and whether we can tell if it does, even in principle, is another question.

Yet another school of thought, occupied by people calling themselves Bayesians, and sometimes called *subjectivism*, takes your chance of getting cancer to be a gaming matter: what are the odds you can get from the bookmakers? What odds would *you* offer? For these are probabilities. If you really must, you can organize a futures market, and sell options on your being diseased. The same for the candidate: let's go to Intrade and see what the market says, let's go to the bookmakers on the high street or in a back alley, and make a bet. Indeed, this sort of behaviour has become rather popular of late for determining probabilities of all sorts. Again, this school would hold the same for our candidate: does spending increase the chance of winning? Let's see what the odds are.

The last school of thought, the *logical* school, looks at probability as a matter of symmetry. This is a familiar notion: when we flip a coin of which we know nothing, what is the probability that it will land heads? They answer: there are two options, and so let us assign probabilities symmetrically. This is a very natural notion, and turns out to be of surprising power. But, it might turn out that we cannot assign a probability to your chance of getting cancer at all using symmetries. If, however, medical science does uncover symmetries, we can then apply the probabilities. The same goes for our candidate: perhaps political science will find a way to describe campaigns in such a way that logical probabilities can be assigned. But if there are no symmetries, there will be no probabilities.

Each school, each interpretation of probability, houses great diversity; they do not speak with a single authoritative voice. Not only are they not univocal, they can be pluralistic, cooperating with each other, combining their answers. This book is a survey of these interpretations, and in it I've aimed to give the main arguments. Chapter 1 is devoted to the relative frequency interpretation, Chapter 2 to the propensity, Chapter 3 to the subjective, Chapters 5 and 6 are devoted to the logical and maximum entropy interpretations, Chapter 4 is devoted to combinations of frequency, propensity, and subjective interpretations. One common theme arising from the choice of the arguments, and my presentation of them, is the fallibility of inductive reasoning. The conclusion that emerges is that the problem is pervasive and unavoidable.

This is not a book about mathematics—it's about philosophy, about the interpretation of mathematics. To discuss the interpretations, at least at the introductory level of this book, we need only basic mathematics (high school algebra and elementary symbolic logic). Any time I go beyond the basic mathematics, I have put it in the appendices, with two exceptions. The first exception is Appendix A.2, which contains a discussion of the basics of probability, namely, its axioms. If you continue reading, you should, at some point, become familiar with the axioms of probability. When the time comes, maybe now, I recommend that you read this appendix. The second exception is Chapter 6. Any attempt to briefly survey the philosophy of probability must include an account of the Maximum Entropy Principle. But while the Principle can be introduced with basic mathematics, it cannot be adequately discussed without somewhat more advanced techniques. I have still tried to keep the technicalities to a minimum, but the chapter is somewhat independent of the others. The remaining appendices provide more technical information about various topics throughout the book: for example, if you need a refresher on logarithms, turn to A.0.5. As to the notation, I sometimes use set-theoretic notation, as in Chapter 5, while in other places, like Chapter 3, I use the logical connectives. This is in keeping with the usual practice of these interpretations of probability, and, for the purposes of this book, harmless.

Acknowledgements

I must single out three friends for special thanks: James Hill, Ondrej Majer, and Peter Milne. James read the entire book in several drafts, and assured me that I could write in my own voice. Ondrej, my constant philosophical companion, was always there to help pull me out whenever I was mired in a philosophical swamp. He carefully read the entire manuscript, and was always willing to listen. Peter has deeply influenced my approach to many, if not most, of the issues in this book. He has also read the entire draft, and never held back. I thank them for their philosophical and moral support.

Petra Ivaničová read the entire Czech translation, found many mistakes, was helpful at every stage of writing, and is an absolute gem. Jaroslav Peregrin cheerfully read each chapter and asked for more, all the while giving me the benefit of his penetrating philosophical insight. Tomasz Placek read the entire book, gave detailed criticisms, and is a great host. Tanweer Ali, my favourite political consultant, read several chapters and still bought me a beer. Vladimír Svoboda went over chapters of the Czech translation with a very fine-toothed comb indeed. Katarzyna Kijania-Placek read a chapter, explained clearly and forcefully why she didn't like it, and throws great parties. Tomáš Kroupa unleashed his formidable philosophical and technical acumen on the manuscript, improving both its content and structure. Gary Kemp gave my explication of Quine the benefit of his scholarship, and has led me to think more deeply about many of the issues in this book. Two anonymous referees provided detailed, constructive, and thoughtful criticism which greatly improved the exposition of many topics herein. Naturally, inclusion in the acknowledgements cannot be taken as an indication of shared blame for infelicities in this book, although it can be taken as having helped to significantly reduce their number.

My views on many issues have been influenced by discussions with friends on quite different topics, although they may not realize it. In particular I'm lucky to know Robin Findlay Hendry, Jonathon Finch, Pavel Materna, Marco del Seta, and Prokop Sousedík. Now is also a good time to note that the characters in this book are fictional, even if they share the names, and even some of the characteristics, of my colleagues.

There are also a number of people who set me on the philosophical path leading to this book. My parents, Clinton and Bettye Childers, raised me to think independently. Hussain Sarkar introduced me to the philosophy of science and formal approaches to philosophy. Donald Gillies and Colin Howson co-taught the first course I took in the philosophy of probability, and have been supportive ever since. Peter Urbach, my doctoral supervisor, was truly a *Doktorvater*. I am deeply grateful to them all.

Work on this book was supported by grants GA401/04/0117 and GAP401/10/1504 of the Grant Agency of the Czech Republic. I am also grateful for the support of the Institute of Philosophy of the Academy of Sciences of the Czech Republic and its director, Pavel Baran, and for the support of Petr Kot'átko, head of the Department of Analytic Philosophy at the Institute.

Finally: Daggi, Laura, and Lukas are my favourite people in the world. I dedicate this book to them.

List of Boxes and Figures

1

Probability and Relative Frequencies

1.1 Introduction

Prokop has secured a scholarship to study in America, and wants to buy car insurance, an insured car being a necessity there. He's 23 and single. Much to his dismay, he learns that he has to pay almost twice as much as a 26-year-old man. The agent explains to him that while 15-to-24-year-old drivers make up 13.2 per cent of the driving population, they are involved in 25 per cent of all fatal accidents. As an age group they are by far the most dangerous drivers. And men are much worse than women within this age group: they are about three times more likely to be the driver in a fatal accident, for example.[1] Since the insurance company is taking a much greater risk in insuring such men against accidents, they will require greater premiums.

Prokop then goes to buy his health insurance, this also being of vital importance in America. He has been known to indulge in various tobacco products, and tells the agent so. Once again, he discovers that he has to pay more for his health insurance: a quick trip around various agents (you can try this online) makes it clear that he will have to pay around 300 per cent more than a non-smoker. When he complains, the broker snickers and says 'Have you ever discussed smoking with your doctor?' In fact, Prokop already knows that smokers as a group have dramatically more cases of cancers and heart diseases than non-smokers.

Prokop will have to spend more on insurance than he planned for. He recalls the sad gambling ('herna' or 'game') bars of his native country, and

[1] NHTSA 2011. Figures 19 and 25 help with visualizing the data.

rules out a visit to the casino to boost his income. Instead, he retires to the library to read about the chances of winning various games in a casino, and wonder when his bank account will be filled by the next instalment of his scholarship.

Prokop's chosen area of study involves a fair amount of physics: his daydreams have been filled with colliding gas molecules. While it may be impossible to predict how single molecules will collide, it is possible to predict how great masses of colliding molecules will behave and hence the behaviour of the gas as a whole. He has learned how the same mathematical techniques used in gas mechanics are also used in a multitude of explanations ranging from the stability of the orbits of planets to the spread of disease and the speed of losses at roulette tables.

Chance phenomena have great importance in Prokop's life, as we can see. Traditionally, the tool for talking about such phenomena is the probability calculus. The purpose of this chapter is to introduce the notion of probability commonly used in insurance, casinos, and celestial mechanics. This will take us first through Richard von Mises's interpretation of probability, and then to various interpretations of A.N. Kolmogorov's axiomatization.

1.2 Von Mises's Relative Frequency Interpretation

The phenomena in Prokop's story are usually held to be described by probabilities termed 'physical' or 'objective'. This is unfortunate, given how philosophically loaded these terms are, and it would be best to avoid them. That, however, is simply not possible, and so we must labour under them. The standard theory of objective probability, as in the theory most likely to be vaguely invoked in introductory texts to probability or statistics, is the relative frequency interpretation. The most popular version of this interpretation (at least among philosophers who are concerned with such matters) seems to be Richard von Mises's. It certainly is the most worked out, and it contains elements found in all relative frequency interpretations in one form or another. It therefore makes a good entry into the field.

1.2.1 Probability and mass phenomena

Relative frequency theories are designed to deal with what von Mises called mass phenomena. These are phenomena (or, better, types of outcomes of observations) that occur in very great quantities. This may be either because we can produce as many as we want by, for example, undertaking an experiment, or because there are a great many instances of them to be found. These events allow us to make 'a practically unlimited sequence of observations' of the same sort of phenomena (von Mises 1957: 11). Von Mises identified three types of mass phenomena: those generated by games of chance (coin tosses, rolls of die, the activities of the gambling hells); 'social mass phenomena' (the subject of life and health insurance—men of a certain age dying, or someone getting a certain illness); and 'mechanical and physical phenomena' (molecules colliding, Brownian motion) (von Mises 1957: 9–10).

Because of its requirement that probability only apply to mass phenomenon, the relative frequency interpretation excludes a number of colloquial uses of 'probability'. One of von Mises's examples was the probability of the declaration of war between Germany and Liberia. Liberia has declared war on Germany in both World Wars (reported in the *New York Times* on 8 August 1917 and 28 January 1944. I don't know if Germany reciprocated). Another example is the probability that a particular candidate will win a US presidential election: no president has ever won an election four times (in fact, only Franklin Delano Roosevelt did so) and unless the twenty-second amendment is repealed, no president will ever serve more than two terms. In general, one-off (or two-off or three-off . . .) events are not mass. Such events take centre stage in Chapter 3.

There is, however, no sharp boundary between mass and non-mass phenomena. A borderline case, according to von Mises, is the question of the reliability of witnesses, presumably in jury trials: 'We classified the reliability and trustworthiness of witnesses and judges as a borderline case since we may feel reasonable doubt whether similar situations occur sufficiently frequently and uniformly for them to be considered as repetitive phenomena' (von Mises 1957: 10). Indeed, there has been much research over the years on the reliability of witnesses, some of it quite frightening (for example, Connors et al. 1996), and it does seem to be amenable to characterization as a mass phenomenon. For von Mises, the boundaries of mass phenomena are determined by actual scientific practice. We shall return to this topic in section 1.3.4, and elsewhere.

Von Mises's interpretation of probability is meant to ground a scientific theory of certain types of phenomena, namely, mass phenomena, and is to stand or fall with the confirmation of its applicability in practice. In this respect von Mises, an avowedly applied mathematician, saw probability theory as parallel to applied geometry and mechanics (see for example his 1957: 7).

Von Mises noted two characteristics of mass phenomena discovered by those who investigate them (actuarial scientists, casino employees, and physicists). The first is that prediction of particular individual outcomes in a series of mass phenomena is impossible. The second is that nevertheless mass phenomena as a whole do exhibit a certain sort of regularity. These he called the empirical axioms of randomness and convergence, and they form the basis of his theory of probability. We will look at convergence first.

1.2.2 Convergence of relative frequency

Consider the humble coin toss.[2] For any given number of flips of the coin, a certain percentage will be heads. We can use the percentage of the occurrence of heads in the total number of tosses as a measure of likeliness. If heads are very likely, then the percentage of heads will be near 100 per cent (nearly 100 out of 100, i.e. 1), that is, heads will turn up at nearly every toss; if they are very unlikely, then the percentage will be nearer 0 (nearer 0 out of 100, i.e. none). If it is just as likely that the coin lands heads as tails (i.e. the coin is *fair*), then the ratio should be one-half. The name for the ratio of heads to total tosses is the *relative frequency* of heads.

Most textbooks of probability assert that the ratios of outcomes to total experiments (involving, of course, mass phenomena) settle down to a definite value. So, when tossing a coin we will observe that the ratio of heads to total tosses converges to some specific value: that is, it settles down to some value and does not deviate very far from this value. Figure 1.1 is a picture of the relative frequency of heads in a series of tosses of a 5 crown coin:

As can be seen, the size of deviations in the relative frequency of heads (or tails) quickly decreases, with the relative frequency settling down, *converging*, to around .49. (The excel file with the data from the coin toss can be found at <http://probability.cz>.) This is a feature of many mass phenomena often observed: we shall take the fact that relative frequencies settle down as one of the bases of our definition of probability. The idea is both common and venerable: von Mises notes that it can be found in Poisson in 1837 (von Mises 1957: 105). Still, more work needs to be done to make that idea mathematically acceptable, and to this we now turn.

[2] In this book, coins don't land on their sides—they land heads or tails. If your coin lands on its side, flip it until you get a heads or tails. If this becomes problematic use a different coin.

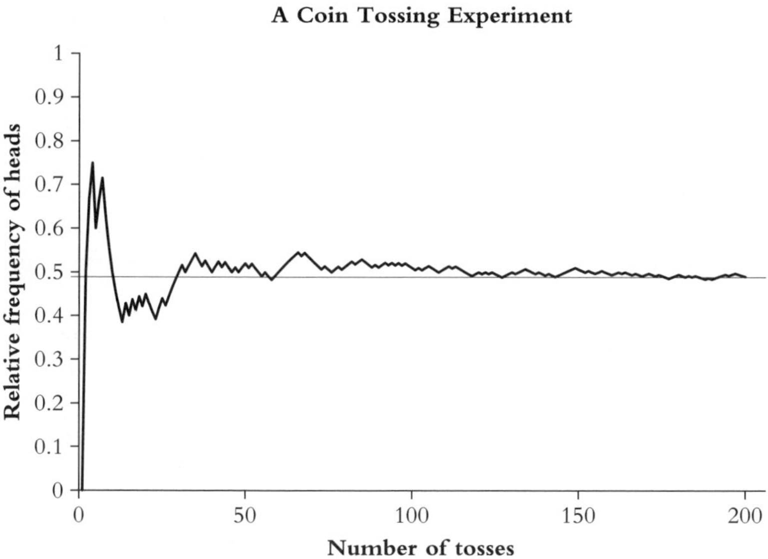

Figure 1.1 A coin-tossing experiment

To introduce the jargon we've already used, the flipping of the coin is variously referred to as a *trial* or *experiment*: this is a general term for phenomena of interest. The result of the trial is called an *event* (or *attribute* or *outcome*). Results of an experiment that we are looking for are called *successes*, otherwise *failures*. So, if we are flipping a coin and are looking for heads, we call the occurrence of a head a success. We could, of course, take tails to be success: nothing hinges on the name. These *elementary* events can be combined into *compound* events, such as two tails occurring in a row, or the tossing of two coins together, or the roll of a die and the toss of a coin.

The set of the names of events to which we wish to assign probabilities is known as the *sample space.*[3] Take for example a roll of the die: we can label the six possible outcomes $1, \ldots, 6$, and so the sample space for the experiment of rolling the die is $\{1, \ldots, 6\}$. For a coin toss the labels are H,T, and the sample space $\{H,T\}$.

When dealing with von Mises's theory we will be only interested in infinitely long repetitions of experiments. An infinite series of rolls of the

[3] Again, nothing turns on the somewhat unfortunate name: see Gillies 2000: 89, for a discussion of the name. Von Mises used the term *Merkmalraum* or 'label space'.

die is represented as an infinite sequence of the integers 1, . . . , 6: say, 1, 5, 3, 2, . . . For coin tosses, the infinite series might look like HTHHTTH . . . The sample space is then the set of all these infinite sequences. A *collective* (in German *Kollektiv*) is a member of the sample space (that is, an infinite sequence of outcomes) that obeys certain restrictions to be introduced shortly. Von Mises stressed its importance in his slogan 'First the collective, then the probability' (1957: 18). As opposed to the approach in 1.3, von Mises defines probability on the basis of the collective, and does not take it to be a primitive notion.

Instead of dealing with events, it is usually much more convenient to work with *random variables* (about which see A.3). A random variable is just a function that attaches a (real) number to members of the sample space. Random variables are very convenient indeed. One of the simplest, and yet most important, uses is that they allow us to attach frequencies to events. In the case of the coin toss, our sample space consists of a series of H's and T's. Turning these into 1's and 0's allows us to add up the amount of heads. Define the random variable I_i so that it takes the value 0 if the ith member of the sequence is T, 1 otherwise. I is an *indicator* variable, indicating if the ith toss was heads or tails. This may seem rather picky, but it is important to keep the representation of our coin-tossing experiment, i.e. a sequence from the sample space, separate from the actual experiment of tossing the coin.

Define yet another such variable S_n to be the sum of $I_1 + \ldots + I_n$. The relative frequency of heads in n tosses can then be nicely written as

$$\frac{S_n}{n}$$

The *limiting* relative frequency of an attribute is the relative frequency as the number of trials grows arbitrarily large, i.e.

$$\lim_{n\to\infty}\frac{S_n}{n}.$$

The first axiom of von Mises's theory is *the axiom of convergence*: for a given collective, the limiting relative frequency actually exists. This means that the value of the relative frequency does not oscillate, but settles down to some value. The axiom is meant to be the mathematical counterpart to the observation that limiting relative frequencies of attributes in mass phenomena settle down. The use of infinity signals the switch to mathematical idealization.

Surprisingly, perhaps, there may be no limiting relative frequency, since the value might just endlessly oscillate. One particularly simple example from Binmore 2009: 104, put nicely in Skyrms 2012, is the relative frequency of natural numbers that begin with 1. This oscillates between 1/2 and 1/10: the peaks and troughs of the relative frequency are 1/9, 10/19, 10/99, 100/199, 100/999, 1000/1999, 1000/9999 . . . (We should note, however, that this example violates the requirement of randomness in the next section.)

1.2.3 Randomness—the impossibility of a gambling system

The notion of limiting relative frequency was not original to von Mises. What was original to his treatment was his notion of randomness. As noted earlier, another key feature of mass phenomena is that it is not possible to predict *individual* outcomes—true mass phenomena are unpredictable. The clearest examples can be found in the gambling hells. A quick search will reveal that there are many betting systems for various casino games and for roulette in particular. A simple example of such a system is: bet on black after the ball has landed on red three times in a row. Following a system like this is a sure road to ruin, or at least to losing at the roulette table, as you can confirm at your local casino.

A change of scenery. Prokop and Horst are on an interminable carriage ride through the nineteenth-century German countryside (don't ask how he got there). Politely declining Prokop's suggestion of a drinking game (the road is too bumpy) and of a sing-along (he's heard Prokop sing before) he instead proposes a game of wagers to while the time away. Horst challenges Prokop to bet whether the next stone marking the distance of their journey (a *Wegestein*) will be larger or smaller than the previous one. But this is a dirty trick, because the road in this remarkably orderly part of Germany has a large stone at every kilometre, with a smaller stone at every tenth of a kilometre (this example is adapted from von Mises 1957: 23). The relative frequency of the larger milestones to total stones will be 1/10, and Horst has made a mental note of where the last one was. He can therefore win every bet. Prokop, finding Horst's idea of amusement decidedly odd, declines and returns to gazing wistfully out of the dirty window. The moral? Horst's game is not random, since it has a gambling system that can guarantee that he wins, and so the sequence of milestones is not a collective.

It is important to note that randomness does not mean fair. Suppose Prokop and Horst take to flipping a coin which, unbeknownst to them, is weighted to heads (it really is a very dull carriage ride). While it lands heads more often than not, they cannot predict when the heads will come up solely on the basis of the results of the previous flips. They cannot, as it were, 'beat the odds' by choosing to bet heads after, say, three tails in a row. It should also be noted that a gambling system need not ensure you win every time. Instead, it should, at least, ensure that you win more often when following the system.

To illustrate how collections of events can fail to be collectives: births in the Czech Republic are recorded at town halls of the local municipalities in large attractive cloth-bound books. Imagine the ministry in charge wants to check the ratio of boys to girls over the last century, but does not want to pay the cost of having someone tabulate 100 years' worth of paper records. Three groups are assigned to the task, but instead of checking each and every record one group each is assigned to check the first entry on each page of the book, the next to check the last entry, and the other to check the entry in the middle on every fifth page. If the register of births is a collective (and this is quite the idealization), then each method will yield the same relative frequency of baby boys (which, most boringly, has been confirmed by digitizing the records and automating the counting), i.e. .53.

How could this list fail to be a collective? Imagine, if you will, that some bureaucrats thought it would be aesthetically pleasing to start each new page with a boy, and end it with a girl. The first group will make the startling discovery that the relative frequency of births of boys is 1, while the second will find the relative frequency of girls is 1. (The third group will also find their numbers skewed. Feel free to explain why.) The data would then not be random in von Mises's sense. Naughty bureaucrats! (And it wouldn't be random even if the bureaucrats *per impossibile* had occasionally slipped up and forgotten to put a boy at the top or a girl at the bottom.) The data could be made into a collective by laboriously snipping up the records, placing them in a large drum, and drawing them out—that is, by randomizing the data.

More prosaically, seasonal events will not yield a collective. For example, the frequency of flowering on a given day in Northern Europe is much higher in spring than in winter. So, picking out days in the spring months gives a higher relative frequency of blooms while picking out days in the winter gives a lower relative frequency. (We could randomize this information by removing data about months, although why we would want to is beyond me.)

Using the terminology developed so far, we can define randomness in terms of attributes in collectives (balls landing red in repeated throws on the roulette wheel, coins landing heads, etc.). The essence of a gambling system is that relative frequencies change when we pick out particular members of the collective (that black is more common following four landings on red, tails more common after five heads). So our first attempt to define randomness is that a collective is random if there is no way to pick out a subsequence of the collective that has a different relative frequency of the relevant attributes than the collective as a whole. Or, put differently, arbitrarily chosen subsequences have the same limiting relative frequency as the sequence.

This simplistic definition of randomness will not do, as the astute reader may have noticed: it allows us too many ways to choose subsequences. For example, suppose I choose a subset of rolls of the die that come up 3. Then the relative frequency of 3's will be 1 in this subset—that's how I picked them. But the relative frequency in the collective can be something different. As well, there will be another subsequence that includes only 6's (and another that includes only 5's, and so on . . .). This can be done with any sequence, and so, given the simplistic definition, there are no random sequences. We must thus allow only ways of choosing subsequences that are independent of knowledge of the outcome. (This argument is sometimes attributed to Kamke, although Hausdorff seems to have made it earlier, albeit in a letter, in January 1920. Föllmer and Küchler 1991: 114.) A standard complaint about von Mises's interpretation was that it was not completely mathematical. And it is true that it took many decades before the notion of randomness was satisfactorily characterized. The next few sections will trace that development.

1.2.3.1 Wald on collectives To make our notion of randomness mathematically tractable, we have to have a function that picks out infinite subsequences of collectives. We will call this function, following von Mises, a place selection function. As we have seen, not just any place selection function will do. First, we need a function that decides for each member of a collective whether it is to be a member of the subsequence independently of what the value of the member happens to be. (For example every fifth member of a collective, or every member that follows four occurrences of some attribute like 'lands on red'.) Such place selection functions are called 'admissible'.

But it also is necessary to show that there are any collectives at all, that is, that there really are such random sequences of attributes. Wald succeeded in showing this in 1937 by proving that if we allow only countably many (see Appendix A.0.3) place selection functions there are continuum many collectives with the proper relative frequency. The natural question is then why should we restrict ourselves to only countably many place selection functions. Wald suggested that we think of gambling systems as theories in a formal logical system (that is, a gambling system would be a set of sentences in a formal language). Since there are only countably many such theories (sets of sentences), this would achieve the desired result of restricting the class of place selection functions. It also seems a reasonable suggestion: surely a gambling system would have to be somehow formalized. Yet, Alonzo Church noted that this suggestion was in fact problematic. Luckily, he also gave a solution to the difficulty.

1.2.3.2 Church's solution In a footnote near the end of Church's 1940 paper 'The concept of a random sequence' he pointed out two problems with defining a theory of randomness in a given formal system *L*; first, the definition of randomness will be relative to that language, and this, to Church, seemed arbitrary; secondly, the notion of defining a gambling system relative to a logic leads to well-known problems:

> It [Wald's interpretation of gambling systems] is unavoidably relative to the choice of the particular system *L* and thus has an element of arbitrariness which is artificial. If used within the system *L*, it requires the presence in *L* of the semantical relation of *denotation* (known to be problematical on account of the Richard paradox). If it is used outside of *L*, it becomes necessary to say more exactly what is meant by 'definable in *L*,' and the questions of consistency and completeness of *L* are likely to be raised in a peculiarly uncomfortable way.
>
> (Church 1940: 135)

Church's insight was that a gambling system could be seen as an *algorithm*. An algorithm is a mechanical step-by-step method for producing some output given some input. Familiar examples are long division and multiplication, and flowcharts are familiar graphical representations of algorithms. We can think of recipes, especially those from *Cook's Illustrated*, as algorithms. Similarly, a gambling system can be thought of as a recipe for success in the casino. More prosaically, a betting system can be seen as a computer programme that tells us whether to bet on a given play on the basis of what happened in previous plays. As Church 1940 puts it, a betting system is an algorithm that computes the values of a place selection function, telling us when and how to bet.

This ties the notion of randomness to a set of ideas centred on the notion of computation: when we compute, we follow an algorithm. If there is an algorithm for a function we say the function is effectively calculable. The notions of 'computability' and 'effective calculability' are the intuitive counterparts of a rich mathematical theory (known variously as computability theory or recursion theory). Perhaps the most well-known part of this theory was given by Turing in 1937, who explicated the notion of computation in terms of a simple abstract machine. A *Turing machine* consists of (1) a finite set of instructions that moves a (2) (re)writing instrument about a (3) tape according to the contents of a (4) memory device containing information about the state the machine is in. This can be moved up a further level: a *universal* Turing machine computes what Turing machines do: it has another computer's action table and input as its input. Thus we can, and will, equate Turing Machines with programmes on a universal Turing machine.

Turing computability has turned out to be a remarkably fruitful notion: algorithms can be thought of as programmes that can be run on Turing machines; effectively calculable functions are those that can be computed by Turing machines (or, if you prefer, on universal Turing machines). The functions that can be calculated by Turing machines turn out to be those known as the (partial) recursive functions, which arose as another means of capturing the notion of algorithm. In fact, it turns out that the proposals for explicating the notion of algorithm that almost all mathematicians find to be acceptable are equivalent. This is why most accept *Church's Thesis*—the claim that Turing computability adequately captures the notion of computability. Thus it is plausible to take an algorithm to be a programme on a Turing machine.

There can only be countably many programmes for a universal Turing machine (the programme must be of finite length, and in a language with a finite alphabet there are only countably many finitely long sequences of letters, and hence only countably many programmes expressible in a given language). We thus find a natural restriction on place selection functions, since it does seem that a gambling system should be algorithmic.[4]

[4] '"When you're operating a system", said Norman, "you can't be gambling. Any hint that you are means that you're not playing the system. That's why to gamblers this approach is boring, pointless, stupid, and takes the fun out of the game. The thrill for them comes in being lucky, while for a system player there is no stroking your rabbit's foot and getting a big kick when you win. Your behavior is completely determined. You should play like an automaton".' (Bass 1991: 42)

To see just how natural this restriction is, consider real numbers. There are uncountably many of them, yet we can only enumerate the values of countably many of them (among these are the familiar π, e, and $\sqrt{2}$). We can only enumerate the values of *computable* reals because they follow patterns. The values of most reals, however, cannot be so enumerated. In other words, there is no Turing machine that has these reals as outputs. One set of real numbers is of particular interest, those between 0 and 1. Each of these reals can be represented by a binary decimal expansion, an infinite string of 0's and 1's. These strings can be taken as possible gambling systems: if the nth place in the string is 1, bet, don't bet otherwise. Only countably many of these gambling systems are computable. So to use a non-computable gambling system would correspond to using the binary expansion of a non-computable real. It is rather difficult to conceive how we might gain access to this expansion.[5]

Gillies 2000: 105–9, who gives a clear account of Church's proposal, regards this as the last word in the explication of randomness. And it does indeed suffice for the purposes of providing a foundation for the relative frequency interpretation. However, von Mises's notion continued to inspire the search for characterizations of randomness.

1.2.3.3 Randomness—Kolmogorov and after Kolmogorov proposed a new definition of randomness in the early 1960s for *finite* sequences of outcomes: the resulting area of mathematics still thrives. The *Kolmogorov complexity* of a string of digits is the length of the shortest computer programme (which we can take to be a programme on a universal Turing machine) that outputs that string of digits.

Consider the small but perfectly formed string 01010101010101010101. In any reasonably flexible programming language there will be many programmes (indeed infinitely many) that print this string. For example, the programme 'print 01010101010101010101'. However, we might consider a programme to the effect 'print 0101010101 twice', or, of course, 'print 01 10 times'. We don't have so many options for a highly irregular

[5] The alternatives to a computable gambling system seem peculiar indeed. For example, if there is a god with access to a non-computable gambling system, and he is inclined to help me, then he might give me access. But then why would I need a gambling system with such powers? Or, I might have a form of precognition (which is what the ability to enumerate the digits of a non-computable real would amount to). Again, though, if I had this ability I wouldn't have to resort to using a gambling system, since, presumably, I would already know the outcome.

string, however, since we can't iterate the print command (for example). This is Kolmogorov's insight: if a string is very disordered, then the length of the programme expressing the string will be at least as long as the string itself.

A caveat is obviously in order: the length of a programme to print a string depends on the resources of the language. So Kolmogorov complexity must be defined relative to a programming language (more accurately, relative to a particular universal Turing machine). But Kolmogorov showed that there is a shortest programme that can express a string, up to some constant, where the constant depends on the language.

This remarkable result means that there is an almost universal measure for complexity. Consider different lengths of 'Hello World' programmes (of which there are many in many different languages). When Prokop was first learning to programme, he used BASIC, where the programme was a simple 'PRINT "HELLO WORLD"'. His friends who went on to become programmers used to hold contests to see who could write the longest non-redundant programmes to print some given output in Postscript. The programmes we are considering are the same: we want to print out a series of digits, and there are more and less efficient ways of doing so. Nonetheless, as the length of the digits printed increases, the portion of the programme dedicated to outputting the digits shrinks as a percentage of the total programme.

Kolmogorov's definition works quite nicely for finite strings, and so it would be even nicer to extend it to infinite ones. One very natural notion is to take randomness as a kind of *incompressibility*. The infinite string 010101 . . . is obviously highly compressible. Conversely, it would seem that a highly irregular string would be incompressible, in the sense that, as we encode more and more of the string, the length of the programme approaches that of the length of the string itself. A natural requirement for the infinite case would be that Kolmogorov complexity settles down to a certain value as more of the string is encoded. As with the requirement that relative frequencies converge, it would be nice if at a certain point complexity would remain within certain bounds.

Alas, this attempt will not work. Surprisingly, it can be shown that no sequences converge in this manner: even highly irregular sequences can exhibit large fluctuations in their complexity. To make a very long story short, it turns out that the problem arises from an unrestricted use of encodings for strings: if encodings are required to be *prefix free*, then there will no longer be large fluctuations in complexity, and Kolmogorov complexity can be extended to the infinite case.

Prefix free codes are codes in which no encoding is a prefix of another. Telephone numbers provide a good example of prefix codes (Downey and Hirschfeldt 2010: 121). When Prokop wants to call his parents, he dials 001 420 555 5555 (actually, he doesn't, but we're following film convention for telephone numbers). 001 puts the system in a state to make an international call, and await a country code, which, for the Czech Republic, is 420. If, however, the international code for, say, Bhutan, were 42, then Prokop would in fact be dialing 0555 5555 in Bhutan. In fact, he would not be able to phone the Czech Republic at all unless some provision were made for looking at the telephone number as a whole and then routing the call. For example, all international codes could be made the same length, but this would result in a prefix-free code again. Or, a special digit could be reserved as a comma that would indicate when a code word ends. This would require looking for the comma, and then executing the code once the end of the number has been found. This is equivalent to the Turing machine moving right and left along its tape. But Turing machines using prefix encodings only move to the right.

Prefix Kolmogorov complexity is an attractive concept. The notion of compressibility is intuitive (even if the requirement of prefix freeness is perhaps not, or at least not until the considerations of Chapter 6). The constant for the additional length provided by a programming language is irrelevant: as the programme length goes to infinity its contribution becomes infinitely small. Even better, incompressibility also turns out to be equivalent to a definition of randomness as typicality, proposed by Martin-Löf 1969. This definition of randomness, very roughly, takes randomness as based on probability: a sequence is random if it can pass all computable tests for conformance with the probability calculus.

We shall leave this topic now, only noting that there seems to be a consensus that this definition of randomness can be used to satisfactorily explicate von Mises's second empirical axiom of randomness. (Eagle 2012 is an elegant discussion of randomness and complexity; advanced reference texts to the field are Li and Vitanyi 1997, and Downey and Hirschfeldt 2010.)

1.2.4 Operations on collectives

Von Mises's theory is not just about what happens in one collective; it wouldn't be of much use then. Instead, the theory is complemented by operations over collectives to produce new collectives. Von Mises laid out four operations on collectives to produce new collectives: selection, mixing, partition, and combination. The first operation, *selection*, is already familiar: it is the production of a new collective by choosing an infinite subsequence by means of a place selection function, say, by choosing every fifth member of the original collective. The new collective has the same attributes as the original. By the axiom of randomness, the probability of the attributes in the new collective is also the same.

The operation of *mixing* corresponds to the union of attributes in the original collective. Von Mises's example (1957: 40): the original collective consists of rolls of a die, the attributes being 1, . . . , 6. The new collective is made by combining 1, 2, 4, and 6 into a new attribute, 'even'; the others, of course, are 'odd'. Or, if you're interested in the evils of tobacco, your original collective might be the attributes 'pipe smoker', 'cigar smoker', 'cigarette smoker', 'snuff taker', 'tobacco chewer', and so on for all tobacco products, as well as the abstainer. A new collective from the original could be 'tobacco user' and 'non-tobacco user'. Or you might want to examine the users of posh tobacco products (fancy cigars and the like), non-posh tobacco products (say, those requiring a receptacle to drool in), and those who use no products at all, along with a catch-all category for the remaining products, if any.

The operation of *partition* forms a new collective by picking out members according to whether they have a given attribute or not. To modify an example from von Mises, five trams run through the Štěpánská tram stop: the 4, 6, 10, 16, and 22. You can see trams from a distance, but can't make out the numbers, only whether the tram has a double-digit number or not. The original collective is just the trams passing through the stop. (Normally, the trams follow schedules, and do not arrive randomly, and so the trams passing through Štěpánská would not form a collective. But today is special: the tram drivers have decided to set out on their route when a ticket bearing the number of their tram is drawn from a drum.) You see in the distance that the tram has a single-digit number, but can't make out which it is: so you form a new collective composed of just the 4 and 6. The new probabilities are just those of the arrival times of the 4 and the 6. For example, assume that all the original trams arrive in equal proportions. Then in the original collective the relative frequency of 4's is 1/5. But in the new collective it is 1/2. Partition captures the notion of *conditional* probability (see 1.3.1.1).

The fourth operation concerns the combination of collectives (and so is rather unsurprisingly known as *combination*). The standard example concerns rolls of dice, but it's very boring. Instead, let's join Prokop, a fan of Bogart, in the back room of Rick's Café Américain. He notes that many people drop by to watch the results at the roulette table. Rick walks in, stands next to a sad-looking young man, who then puts all his remaining money on 22. The ball lands on the 22, and, with Rick leaning over his shoulder, the young man repeats the gamble on 22, where the ball lands again. After a brief exchange with Rick the young man hastily leaves the room, and Prokop's eyes mist up.

Consider two collectives, one consisting of the people standing next to a certain chair by the roulette wheel, the other consisting of the outcomes of the spin of the wheel. We can combine these to create a new, two-dimensional collective. This new collective can be used to think about sampling from one collective on the basis of another. For example, we might be interested in 22 coming up and Rick standing next to the table. So, we can partition the new collective and look for the occurrence of Rick given that 22 comes up. Some symbols should help. Denote by $rf(\text{Rick})$ and $rf(22)$ the limiting relative frequencies of Rick and of 22 in their original collectives, and let $rf'(22|\text{Rick})$ be the relative frequency of 22 after partitioning the new collective by Rick's presence. Then the limiting relative frequency that 22 comes up when Rick is next to the table is:

$$rf'(22|\text{Rick}) = rf'(22 \cap \text{Rick})/rf(\text{Rick}),$$

or, very slightly rewriting:

$$rf'(22|\text{Rick})rf(\text{Rick}) = rf'(\text{Rick} \cap 22)$$

Now, suppose that Rick never intervenes, his heart never softens, and his presence has nothing to do with 22 occurring. Then the partitioning of the collective will have no effect on the relative frequency of 22, and we would have $rf'(22|\text{Rick}) = rf'(22)$, that is, Rick and 22 would be *independent*, and

$$rf'(22)rf(\text{Rick}) = rf'(\text{Rick} \cap 22)$$

In general, and in this case in particular, this does not hold. The notion of independence (or a lack thereof) is central to probability (as we shall see in 1.3.1.2).

This completes our exposition of von Mises's system. We now turn to some standard objections to it.

1.2.5 *Objections to von Mises's Interpretation*

In this section we will discuss some objections to von Mises's interpretation. Some objections are specific to his interpretation, and others apply to the frequency interpretations discussed in the following sections. I will not provide an exhaustive catalogue, but just try to hit the high points. The interested reader should consult Hájek 2009, who provides fifteen arguments against various versions of frequentism.[6]

1.2.5.1 Ville's objection(s) In 1939 Jean Ville showed that there is a collective of 0's and 1's that has limiting relative frequency 1/2, but that has, for any *finite* initial segment limiting relative frequency greater than 1/2. Such a sequence is 'biased' in the finite initial segments of the sequence. A gambler who always bets on 1 will always come out ahead in any finite amount of initial bets. Maurice Fréchet, Ville's teacher, saw this as a death blow to von Mises's system:

> [Ville's result] is then quite sufficient to show that with their definitions, De Mises [sic] and Wald not only have failed to eliminate all regularities but even have not succeeded in eliminating one of the most easily recognizable of them.
>
> (Fréchet 1939: 21–2)

Fréchet has been followed by almost all writers on the topic of randomness in proclaiming von Mises's account of randomness (and, apparently, his interpretation of probability) as a complete failure.

There are two objections to von Mises's interpretation that arise from Ville's work. First, it seems that von Mises has failed to capture the intuitive notion of randomness. Second, it seems that von Mises has failed to produce the standard mathematical theory of probability, i.e. Kolmogorov's measure-theoretic axiomatization (for which see 1.3 and A.2). In particular, the collective produced by Ville violates certain theorems concerned with deviations of relative frequencies from the mean (in particular, the Law of the Iterated Logarithm, for which see A.5.2). The collective behaves too regularly, and in doing so seems to violate these theorems.

[6] One reason I will not discuss Hájek's objections in detail is that it would require quite a detour through other areas of philosophy dealing with, for example, idealizations in science.

This second is a very serious charge indeed. If von Mises has failed to ground the usual axioms of probability, most will find that far too high a price to pay.[7]

The first objection is less powerful than it seems. Following standard mathematical practice, von Mises used convergence in the infinite limit to obtain a smooth mathematical theory: the result that finite gains are always possible does not say anything about infinite gains. Von Mises is interested in such infinite gains because he wishes to model probabilistic phenomena. If his theory of randomness allows this by grounding the usual axiomatic approach, then he has, by his lights, succeeded.

However, it could be objected that sequences that approach the limit unilaterally really would allow a gambling system, at least in an intuitive sense, and so the von Mises (and Wald and Church) definition of randomness fails to capture the notion of the impossibility of a gambling system. This complaint is made by Shafer and Vovk: 'Von Mises ... never conceded Ville's point. Apparently von Mises's appeal to the impossibility of a gambling system had only been rhetorical. For him, frequency, not the impossibility of a gambling system, always remained the irreducible empirical core of the idea of probability' (Shafer and Vovk 2001: 49).

But rhetoric aside, von Mises's characterization of randomness does ensure that the appropriate relative frequencies in collectives are in fact probabilities. The question of the importance of capturing the notion of a gambling system in a stronger fashion therefore seems somewhat moot. (Howson and Urbach 1993: 324, make a similar point.)

The second objection is that von Mises has failed to provide such a grounding for the probability calculus, since sequences like Ville's violate a law of probability known as the Law of the Iterated Logarithm. This law (which we need not go into detail about) puts limits on fluctuations of sequences, and so excludes such gains in finite cases. It seems that Ville has shown that von Mises's interpretation allows them. This, however, is simply wrong. The sequence that Ville describes has measure zero with respect to von Mises's system and so can, from the point of view of the system, be ignored. (See Wald 1938, von Mises 1938. An explanation of

[7] Interestingly enough, the usual practice is to grant the success of the usual axiomatization, and aim to ground it. Searches for different axiomatizations which yield different results have not gained any popularity, and so it seems that naturalism reigns in the philosophy of probability.

measure zero can be found in 1.3.2.1 and A.2.2.1.) In fact, the Law of the Iterated Logarithm can be proved in von Mises's interpretation, since the interpretation does yield the standard axiomatic framework (see, for example, the appendix of Geiringer 1969, as well as Lambalgen 1987a and 1996). It is worth noting that the same objection could also be made to the standard approach covered in section 1.3. In that account the sample space includes all sequences which a fortiori contains sequences like Ville's. So the mere existence of such sequences alone cannot be the problem with Ville's results.[8]

1.2.5.2 Elegance (or the lack thereof) Von Mises's first attempt at defining probability was in 1919: it was a time of increasing abstraction, and mathematicians were busy axiomatizing. For example, in the last half of the nineteenth century, geometry was viewed as being freed from its messy roots in physical intuition. Probability followed much later. Kolmogorov is viewed as having provided the definitive axiomatization in 1933. This axiomatization showed how to include probability in measure theory, a part of analysis, and this is how it is viewed by most mathematicians today. Von Mises on the other hand viewed probability as a mathematized science. For him probability was a theory of certain phenomena, and this theory had to have some connection with reality. This, to mathematicians at the time, and perhaps now, seemed to have got the matter exactly backwards. Indeed, a common complaint was that von Mises mixed the empirical and the mathematical. (An interesting insight into the tensions between the mathematical and the applied mathematical views on probability can be found in D.V.C. Lindley's 1953 review of J.L. Doob's classic *Stochastic Processes*: 'This is a book of mathematically interesting theory, not a book of solutions to problems . . . [W]ithout some other material the book is rather uninteresting unless mathematical manipulation for its own sake exerts enough of an appeal.' Lindley 1953: 455–6.)

It is, however, almost universally agreed that it is easier to work within Kolmogorov's formulation of probability theory. Doob, in a debate with von Mises (von Mises and Doob 1941), refered to von Mises's approach as 'awkward', 'inflexible', and 'clumsy'. Doob saw von Mises as providing an interpretation of probability (of which he approved) that could be separated from the axiomatization (of which he did not approve, preferring

[8] The discussion of the issues in this section arises from joint work with Peter Milne.

Kolmogorov's axiomatization). This seems to echo Kolmogorov's view of the matter: he refers to von Mises as giving the interpretation of his axioms in terms of relative frequency (Kolmogorov 1933, as well as 1963).

Another way to put it is, as pointed out by van Lambalgen (1987b: 16–17) and by Doob, that von Mises is able to define probabilities from his axioms whereas Kolmogorov assumes them. Hence, one can be taken as an attempt to explain probabilities, the other as an attempt to provide an abstract mathematical structure of great generality deemed useful in making precise some things said of probabilities. Von Mises rejected this characterization of the difference between the two, presumably because of his conception of applied mathematics, which certainly seems to arise from his rather strict operationalism (which is thoroughly discussed in Gillies 1973 and 2000).

1.2.5.3 Infinite limits and empirical content The astute reader may have noticed that for von Mises probability is defined only in an infinite limit. This seems to put serious constraints on the applicability of the system. Not only are research budgets not large enough to observe such sequences, we do not live long enough, even as a species, to observe such sequences. And even if we could observe an infinite sequence, the coin would wear out after some point.[9]

To put it baldly, the problem is that since we can only observe the initial segment of an infinite sequence of events, we can never be sure what the true probability of the event is. For example, the first 1,000 tosses of a coin may have relative frequency .6, then for the next 100,000 relative frequency .4, and all the while the limiting relative frequency might be .5. Nothing in the mathematics of the theory prevents this. Hence, von Mises's theory has no empirical implications, and so seems worthless for application. This is surely a bad result for a theory that is meant to be a scientific theory of the real world.

There are a number of responses to this charge. The first is that limits of infinite processes occur in many places in the sciences, and that these sciences are successful. For example, physics uses the differential calculus, where a derivative is defined as a limit. So, acceleration is defined as the second derivative of displacement, which means that it is also defined

[9] Can an infinite sequence of events occur in finite time? This is the question of supertasks, which arises when considering Zeno's paradoxes: see Laraudogoitia 2011.

using infinities. However, as Howson and Urbach (1993: 335) point out, determining acceleration by taking a limit results in a definite quantity, which can then be checked against an observation. With probabilities, the situation is different, because any finite observations will be consistent with any probability value (excepting the extremes of 0 and 1).

Gillies (2000: 101–3) points out that von Mises could counter that in fact many derivative values in physics are approximations, and in principle do not settle down to a single value. For example, quantum phenomena may cause the mathematics of infinite limits to cease to correspond to the actual physical situation: yet the mathematics remains useful. Thus, observation and theory are not very tightly tied together, but are tied together enough to be useful. Still, Gillies points out (following de Finetti) that the difference between the two cases is that the link between theory and observation in physical cases is due to precisely the notion of approximation, while in the case of probability there is no notion of approximation: the differences between observation and theory are a necessary part of the theory. It's not that observed probabilities approximate true probabilities: they never do so because of the infinity of the sequences used to define relative frequencies.

There are two possible responses to this difficulty: first, it might be argued that probabilities do in fact settle down in most cases. This, as we have seen, seems to be true. But it is not enough to ground a scientific theory of mass phenomena: we need to have some idea of when the limits will settle down. Second, it might be argued that Laws of Large Numbers (for which see A.5), which are theorems about the long-run behaviours of random variables, show that the limits will settle down quickly. They do not, however, show this. The Laws of Large Numbers show that 'most' sequences settle down. But we never know if the sequence we are dealing with is one of these sequences that settle down, or not. (Von Mises did not hold that the Laws of Large Numbers showed that relative frequencies quickly settle down to the mean: he regarded such Laws either as number-theoretic results or as statements of an empirical observation, discussed earlier, of the convergence of relative frequencies.)

Von Mises also pointed out that there is a problem for all scientific theories that is even more general than the use of infinite limits: some form of infinity shows up everywhere. For example (following von Mises 1957: 84–5) determining, say, the specific gravity of a substance would seem to require that we test all of the substance (we can't say the substance always has a certain specific gravity if we haven't tested all of the substance). To use another very well-worn example, we can't tell if all ravens are black until we've seen them all.[10] This is possible, since there are, maybe, only finitely many ravens. But what about the properties of sand? Or of water molecules? If we don't examine them all, can we say that we know what properties sand or water possess? This does not mean that science is wrong: it means that we need a theory of how to link finite observations with theories. This is the key to addressing this particular concern: von Mises would separate his theory meant to describe mass phenomena from a theory of inference concerning which models of the mass phenomena are correct. Theories that address some of these concerns will be discussed in later chapters (von Mises himself was an advocate of a version of the Bayesian account of statistical inference).

1.2.5.4 Single-case probabilities and the reference class Let's return to Prokop, as he munches a ham sandwich (white bread, bright yellow mustard) nicely provided by his elderly neighbour. Thoughtfully unsticking the bread from the roof of his mouth with a post-alveolar click, he wonders how the probabilities of the insurance agent apply to him. Yes, he is an unmarried, young male, who, while enjoying a meaty diet also exercises a lot, and doesn't have much stress in his life. Prokop is an individual. His bright pink leather shoes, his girlfriend-irritating cautious driving style, his wide-ranging intellectual interests, his fondness for lizards, are all part of a nearly infinitely long list of his characteristics. But can we then imagine this list being instantiated infinitely often, infinitely many Prokops living their lives and wrecking their cars and getting lung cancer, or not, as the case may be? The answer is clearly no—there is only one Prokop. We could search for people similar enough to Prokop; in fact, that is what we do. But to get at a chance of Prokop getting cancer, we need to repeat him. Make him into the collective, and not the others. This is why for von Mises probability (of the frequency sort) is necessarily about groups.

But what then should an individual do? Should Prokop quit smoking and eat less meat? For von Mises, this is a question for the science of how to make decisions. We shall discuss an attempt at such a science in section 3.6, but the theory concerns individual choice, not reasoning from a collective.

[10] By the way, not every raven is completely black, since there are cases of albino and leucistic ravens. There are also at least two species of raven, the white-necked raven, *Corvus albicollis*, and the Chihuahuan Raven, *Corvus cryptoleucus*, with white spots.

Once again, let's visit Prokop, as he washes the sandwich down with a can of (remarkably cheap yet tasteless) beer. He reflects on the unfairness of life: he is in fact a very careful and safe driver, much to the irritation of his ex-girlfriend. And even though he smokes, he remembers that there are many smokers who never get cancer (the example that is always trotted out is Winston Churchill, a great smoker and drinker, who lived to be 91). And indeed, sometimes a gambler does walk away with a lot of money. So why not him? Probabilities are, according to the insurance agent, properties of mass events. But what do these mass events have to do with *him*? Von Mises would answer: Nothing. These events have nothing to do with Prokop as an individual. This is known as the problem of the reference class: if we specify an individual, there will be no collective—so how do we specify the collective?

There are two problems arising from the reference class: first, we now need a way to find out how to pick out an appropriate reference class to put an individual in, and secondly, there remains a nagging intuition that probability should apply to individuals.

To the first problem von Mises and his supporters could retort that this is a problem for a different theory: when science picks out the correct collectives in say, mass social phenomena, then his interpretation of probability applies. But the question of going about picking out individuals to put in a collective belongs elsewhere: in the case of cancer, biology should specify the reference class. And, as in the problem of the preceding section, assessing the correctness of a classification is for a theory of inference, and not for a theory of mass phenomena (remember: first the collective, then the probability).

And as to the second problem, the retort would be that if you want to run a successful insurance firm, if you want your casino to make money, if you want your physics to correctly describe a gas, you are not interested in individuals: you are interested in groups. The theory therefore does not, nor is it meant to, assign probabilities to single events like a coin's coming up heads on a given toss or of a certain man dying of a heart attack when he is forty. For von Mises, probability attaches to *classes* of events: the probability of a particular man in his forties dying of a heart attack does not, for von Mises, exist. Instead, we can have the probability of men in their forties dying from heart attacks. Using such information we can set the price of an insurance policy for individual men, so the theory has practical consequences for single men, even if it is defined for mass events. So, when we talk of the probability of heads, we mean the probability of heads in a long series of tosses.

In the next chapter we will consider an interpretation of probability that aims to establish single-case probabilities, and its relation to von Mises's theory. But for now, it is time to turn to the relative frequency interpretation hinted at in many textbooks, one based more directly on Kolmogorov's axioms.

1.3 Kolmogorov and Relative Frequencies

Von Mises's interpretation is a 'bottom up' account of probability: first the phenomena, then their abstract representation in a collective, and then the probabilities. Most textbooks take a 'top down' approach: the abstract theory is introduced, and then interpreted as representations of phenomena of interest. The canonical abstract theory was provided by Kolmogorov in 1933. In this framework probability is characterized as a measure of a special kind: a measure with a notion of independence. As suggestive as the names of the theory are ('random variable', 'expectation'), these quantities remain uninterpreted. However, there is a natural interpretation of the axioms in terms of relative frequencies. In the next section we will give an informal explication of this connection, while in the later sections we will examine some specific versions.

1.3.1 Relative frequencies as probabilities—the Kolmogorov axioms

We begin with an intuitive explication of the axioms, and then to a slightly more rigorous explication of the measure-theoretic framework. In the next section we will show how relative frequencies satisfy a version of the Kolmogorov axioms. This serves to motivate the use of the measure-theoretic framework to explicate frequentism. We approach this in two steps. We give an uninterpreted abstract account, and then describe Doob's reinterpretation of von Mises's interpretation in this account. In the following section we discuss van Fraassen's modal frequency account before turning to the problems of the frequency interpretation.

Relative frequencies are just relative counts of occurrences of events. The set of events we're interested in is, as in section 1.2.2, the sample space. Call the sample space, as always, Ω, and let A be a member of Ω. Let $n(A)$ be the number of times that A occurs in n trials. The first axiom is that the probability of an event is a number greater than or equal to 0. That is, for any event A in the sample space,

(1) $p(A) \geq 0,$

Obviously, $n(A)/n$ trivially satisfies this requirement, for any n. $p(A)$ could be 0: suppose that A means 'the coin comes up heads'. If the coin never comes up heads then $p(A) = 0$, but it can never be less than 0. Thus relative frequencies trivially obey the first axiom of the probability calculus.

The second axiom is that the probability of the certain event is equal to 1. In this case, 'the certain event' just means the entire sample space, since some event in the sample space must happen.

(2) $p(\Omega) = 1.$

Relative frequencies also obey this axiom. In our simplest example, $\Omega = \{H, T\}$. Now, the coin must come up either heads or tails, that is, some member of Ω must occur. Returning to our earlier notation, $n(\Omega)/n = 1$, obviously, since any occurrence of anything will count, and so $n(\Omega) = n$ for all n, and hence, $p(\Omega) = 1$.

The first two axioms are the simplest. They determine that a probability is always between 1 and 0. The third axiom is that probabilities add in a certain way, and this is what gives probability its special character. Consider two events that cannot occur together. For example, a roll of the die can come up either 1 or 6, but not both. Suppose that the die is fair, that is, after repeated rolls, $n(1)/n = n(6)/n = 1/6$. Then the probability that either 1 or 6 will come up should be higher, there being more events to be counted in the numerator. The number of times the die comes up 1 or 6, $n(1 \text{ or } 6)$, is, of course, the number of times the die comes up 1 plus the number of times the die comes up 6, $n(1) + n(6)$. Therefore, the relative frequency is $(n(1) + n(6))/n$. This reasoning leads us to the third axiom that the probabilities of exclusive events add:

(3) $p(A \cup B) = p(A) + p(B), \text{if } A \cap B = \varnothing.$

Most generally, the third axiom says that the probability of any finite collection of mutually exclusive events adds. There is a question, which we will not pursue, of whether the probabilities of infinite collections of (mutually exclusive) events add, that is, if the third axiom should be over arbitrary countable sets (see Appendix A.2.4.1).

The intuitive argument we have given will not work for infinite limiting relative frequencies for technical reasons which we shall not go into (see van Fraassen 1980: 183–7 for an account of these difficulties). Since we only aim to give a motivation for interpreting the Kolmogorov axioms in terms of relative frequencies, these difficulties need not detain us. However, in the infinite case, not all limiting relative frequencies are probabilities, and a formal theory must take care to exclude these problematic cases.

1.3.1.1 Frequentist conditional probability We have so far described the axioms for what is called unconditional probability. But one of the most important aspects of probability is conditional probability. We can illustrate this notion from a frequentist point of view using an example of Feller's (1957: 104–5). Suppose that a scientist is interested in the proportion of a population afflicted by colour blindness. Under the interpretation we have been discussing, the probability of colour blindness is n(colour blind)/n. But, perhaps we are interested in checking the probability of colour blindness in females. A natural way to define this particular probability would be to restrict ourselves to the subpopulation of females, and then check how many colour-blind women there are. In other words, we would look at n(Women and colour-blind)/n(women). This is the same as looking at the intersection of the two sets from our population, women and the colour-blind: n(Women $\cap$ colour-blind)/n(Women). The meaning of this probability could be restated as the probability of someone's being colour-blind given that they are a woman.

More generally, the occurrence of an attribute A given the occurrence of another attribute is $p(A|B)$, and we have the following definition:

(4) $$p(A|B) = p(A \cap B)/p(B), \text{provided } p(B) \neq 0.$$

We shall treat this axiom as defining conditional probability. We shall encounter conditional probability again in other chapters where it will have quite different meanings.

1.3.1.2 Independence One of the most important notions in probability, and one that is essential for an adequate understanding of the relative frequency view of probability, is that of *independence*. Two trials that have no effect on each other are said to be independent. Suppose, for example, that you are tossing two coins and the tosses in no way influence each other. One way to define independence is to say that conditioning one on the other does not change the outcome, that is, A and B are independent if

(5) $$p(A|B) = p(A).$$

Looked at from the point of view of relative frequencies, this means that looking at subpopulations of B where A has occurred does not change the relative frequency. We can reuse the previous example concerning colour blindness. Suppose that, unlike in the real world, gender and colour blindness are unrelated, in that the chance of being colour-blind is the same for both men and women. Then we would expect that if we restrict our attention to women, B, then the frequency of colour blindness, A, would be the same.

Independence can be used as an exact way to capture the intuitive notion of independent trials: equation (5) says that the outcome of one experiment, that is, the presence of one attribute, B, has no effect on the outcome of the other experiment, that is, the presence of A. Trials for which (5) holds are said to be independent. (It is easy to see that (5) can be rewritten as $p(A \cap B) = p(A)p(B)$, that is, probabilities of independent events multiply.) In von Mises's account, independence is, of course, a derived notion. This leads to a different role for its use in von Mises's system, a difference worth noting.

1.3.2 The measure-theoretic framework

The measure-theoretic approach is so named because it investigates generalizations of our familiar notions of measure. This need not detain us now, although sometimes geometric intuitions will help with understanding certain results. As usual, we start with the sample space Ω. Another set known as a field, which we will denote unimaginatively as F, contains all events of interest. It is constructed as follows:

(a) Ω is in F;
(b) if A is in F, so is its complement, A^c;
(c) if A and B are in F, so is their union $A \cup B$.

The last two restrictions ensure that the field is closed under union and complementation.[11] The elements of Ω are the elementary events, while the other members of the field are the events. In what follows we will take it as understood that events A and B are members of the field.

Now for the axioms governing the probability function, of which there are, traditionally, three. The probability function is defined over members of the field as follows.

First, probability measures are non-negative:

[11] There is in general more than one field, and there is always at least one field, consisting of the sample space and the empty set.

(1) $p(A) \geq 0,$

Secondly, the probability measure of the sample space is unity:

(2) $p(\Omega) = 1.$

Thirdly, probability measures of disjoint sets add:

(3) $p(A \cup B) = p(A) + p(B), A \cap B = \emptyset.$

From such humble beginnings the whole of the edifice of the mathematical theory of probability can be built. Note that the theory so far is just about sets and functions which assign numbers to those sets. (We also give a simpler version than usual. (3) is usually extended to cover countable combinations of events; see A.2.4.1.)

We now have a trio, known as a probability space, of sample space, field, and probability measure. This is the foundation of the measure-theoretic approach: a basic set is specified, from this a set of sets is formed, and to this set of sets the probabilities are applied.

1.3.2.1 Measure zero One very useful aspect of this approach is the account of sets of measure 0. These are simply sets that have probability 0 (the probability function is a measure, hence the name). In particular, if a field has uncountably many members, then any set with only countably many members will have measure 0 with respect to that field. (It might help to keep a geometric intuition in mind: relative to a plane, a line has area 0. The line may be infinitely long, but it's negligible in terms of area.) If a set has measure 0, then from the point of view of the measure it can (indeed must be) ignored. Probabilistic laws hold with measure 1, meaning that they hold for 'most' sets. We will meet with the uses of this characterization in the next section.

1.3.3 Doob's reinterpretation of von Mises

Doob 1941 provided a reinterpretation of von Mises's account within the measure theoretic framework. This section provides a simplified account for binomial trials, although this restriction is unnecessary. In this interpretation, the sample space is the set of infinite sequences of 0's and 1's, representing the outcomes of trials. We assign probabilities to the members of the sequences by setting $p(1) = r$ (which implies that $p(0) = 1 - r$), and require that the probabilities multiply. This provides a model of an infinite sequence of independent identical trials. The sample space is uncountably large, and so any countably large member set will have measure 0.

Doob took two theorems to be of central importance in determining that measure theory was the correct abstract account of probability. The first theorem is that the set of sequences of the sample space for which the relative frequencies of the 1's do *not* converge has measure 0, meaning that the size of this set, relative to the size of the sets that do converge, is, to borrow a term from Billingsley 1995, 'neglible'. The second theorem is that the relative frequency of infinite subsequences is the same as that of the original sequence for all but a set of sequences of measure 0. In other words, non-random sequences have measure 0. Thus Doob showed that von Mises's axioms of convergence and randomness can be deduced from the usual measure-theoretic framework. (Feller 1957: chapter VIII provides a good explication of Doob's approach; Billingsley 1995: 27–30 gives an incisive account.)

The measure-theoretic approach interpreted as repeated identical independent trials is, it seems, the standard textbook interpretation, at least when books on the probability calculus provide any interpretation at all. The reason for its popularity is clear: it is mathematically much more elegant.

We can now see how different von Mises's notion is than the one found in many textbooks, namely, that of probability as the relative frequency of outcomes of repeated *independent* trials. This approach uses the notion of independence as a replacement of randomness. Note that in this approach the probability is assumed, and the properties of the infinite sequence are deduced. As Kolmogorov put it, this is opposed to an approach 'in which the concept of probability is not treated as one of the basic concepts, but is itself expressed by means of other concepts' (1933: 2).

1.3.4 Van Fraassen's modal frequency interpretation

Van Fraassen's modal frequency interpretation is formally similar to the textbook interpretation, but is motivated differently. Van Fraassen (1980, section 4.4) takes as his starting point problems with Hans Reichenbach's relative frequency interpretation (Reichenbach 1949. For an explication see Galavotti 2005). Reichenbach's interpretation was less restrictive than von Mises's in that he did not require sequences to be random. As van Fraassen points out, this leads to serious technical difficulties (of the sort encountered by the naive relative frequency interpretation in 1.3.1).

However, he wants to retain the character of Reichenbach's approach by giving an interpretation that takes into account actual observations of frequencies. Such a frequency interpretation would fit nicely with van Fraassen's aim to give an account of how to 'save the phenomena' to complete his particular empiricist account of science.

The basic idea is simple: probability begins with (a model of) actual events, which are then extended to infinite sequences of copies of these events. We can easily do this by considering duplicates of experiments. If our actual experiment is the toss of a coin, then the outcome is described by the sample space $\{\text{Heads}, \text{Tails}\}$, or, more simply, $\{0, 1\}$. Repetitions are represented in the natural way: $\{11, 01, 01, 00\}$, for two $\{111, 110, 101, 100, 011, 010, 001, 001\}$, and so on. These sets are Cartesian products. Infinite repetitions of the experiment correspond to the infinite Cartesian product. Our sample space consists of all these infinite Cartesian products, which is simply the sample space of the textbook approach, and so this approach can then employ the usual machinery of the probability calculus.

The 'modal' comes from constructing a model of possible extensions of the actual events (where the model is a logical model). The modal frequency interpretation has a much closer fit with the textbook interpretation, providing a natural way to interpret the extension of the original sample space by Cartesian products: the initial outcomes are what the possible infinite sequences are built from.

1.3.5 Problems with Kolmogorovian interpretations

Kolmogorovian accounts, while mathematically more elegant, share the same problems as von Mises's interpretation as concerns empirical import in 1.2.5.3 (an issue van Fraassen directly addresses in his 1980, section 4.5) and with respect to the reference class and single-case probabilities (1.2.5.4). They also have some difficulties of their own, namely the use of independence to define probabilities.

The first difficulty for Kolmogorovian interpretations is that circularity looms: probability is explained in terms of trials with constant probability, which then yield relative frequencies. But the relative frequencies are the result of the constant probability, which is, usually, only explained as relative frequencies. We could take the approach outlined in the next chapter, that is, the propensity approach, to overcome this circularity. Most textbooks don't take this route, and, as we shall see in the next chapter, with good reason.

The second difficulty is that of independence. Independence is necessary for the idea of repeated trials with constant probability. But independence remains a magical notion unless specified: and the notion seems to be one of *causal* independence. So such interpretations rely on an account of causation, and this is a notoriously difficult notion. Even worse, it seems doubtful that any successful account of causation will be independent of some theory of probability, since that seems how we must explain probabilistic causation. (Howson discusses the issue of independence and von Mises's framework in section 3.1.1 of Howson 1995.)

1.4 Finite Frequency Interpretations

It might be thought that the problems of the limiting relative frequency approach might be solved, or at least mitigated, by avoiding infinities, and sticking to a finite frequency approach which defines the probability of an event as the actual relative frequency of its occurrence. (Such interpretations are sometimes termed 'actual' frequency approaches as opposed to 'hypothetical' frequency approaches which employ the idealization of an infinite series of outcomes.) I can find no clear example of anyone holding such a view (although Hájek 1997 disagrees). Such an approach would face severe difficulties, which are laid out by Hájek. For example, a misleading initial sample could give a very different probability of an attribute than the actual relative frequency of the attribute in the population.

However, Peter Milne has pointed out to me that there is a sense in which finite relative frequencies are perfectly respectable probabilities. For example, consider a sample space that consists of all species of fish in a particular lake, say, Máchovo Jezero. We can assign probabilities to the occurrence of a random sampling of a given species by assigning the actual percentage of that species in the lake. So, if 75 per cent of the fish are striped bass, then the probability of the occurrence of a striped bass would be .75 under certain assumptions. We might not know the actual frequency, but that's a matter for statistical inference (for example of the kind discussed in Chapter 3). This probability obeys the axiom, and presumably could be useful in certain circumstances—other examples might be voting preferences or intentions to buy particular consumer products. Still, such finite frequencies would not be about mass phenomena, and so would not be a substitute for the limiting relative frequency interpretations or the propensity interpretations covered in the next chapter.

1.5 Conclusion

This chapter has presented two families of relative frequency interpretations. It is clear that they share the same prospects and problems. In particular, both interpretations are in a sense incomplete, in that they need to be supplemented with a theory to link theory and evidence, both in the case of relating a series of observations to a collective and in linking probabilities to individuals. Von Mises was well aware of this: his theory was not meant to provide an account of how we can tell if a scientific theory, including his, is properly applied, or even true. (This is stressed by Gillies 2000: 99, who points to the preface to the third German edition of von Mises's book.) Subsequent chapters will address some suggestions for how to provide the necessary links, and in particular the next chapter will address a different interpretation of probabilities meant to assign probabilities to individuals.

2

Propensities and Other Physical Probabilities

Prokop goes back to his insurance agent. 'Sure, most 24-year-old single men drive like they're mad. But *I* don't! Why should *I* have to pay more?' Indeed. What does the fact that most young men drive like maniacs tell us about Prokop's driving? It turns out that the insurance agent has a degree in philosophy: he patiently explains that the company has collected data which shows convincingly that young single men get into more accidents. Prokop cries 'That's unfair! I drive more carefully than my grandmother'. The insurance agent shrugs his shoulders and says 'How can we know that? We just know about the group you belong to, and that group tend to drive like maniacs'. Prokop now knows that it's not nice to be picked out for a penalty just because you belong to a group, membership of which is not voluntary.

Prokop has the urge to point out that he is Czech, and that there were fewer accidents per capita in the Czech Republic, at least in 2004, than in America (Economic Commission for Europe 2007: 8–9). But he also doesn't feel like getting drawn into a long actuarial discussion of various reporting systems of possible road safety issues that might lead to accidents, for data does not come neatly organized, even within the European Union (Vis and Van Gent 2007). He does bring up that even though insurance companies charge unmarried men more than married men, some studies have shown single men to be involved in fewer fatal accidents (Kposowa and Adams 1998). Finally, he does not reveal that his grandmother's nickname is Two Wheels Marta, for the number of wheels firmly on the ground when she went around a corner; Prokop's claim to drive more safely than her is therefore somewhat misleading.

The agent regards him curiously. Prokop is convinced that if the insurance agent knew everything about him, he would know that he, Prokop, is a safer bet than other single men his age—and what is insurance other than a bet that he won't get into an accident? Prokop believes, rightly so, that he's unique, that he really does belong to a reference class of one (section 1.3.4). And he believes that there really is a chance, an objective chance, of his getting into an accident and that it is less than his grandmother's chance of getting into an accident. Moreover, he thinks that there's a chance he will get into a wreck each time he drives, small as that chance may be. Prokop thinks, in other words, that probabilities need not apply only to repetitive classes of events, but to individual events as well. The view that probability attaches to individual trials rather than collectives or the like is decidedly not a frequentist view of chance. We'll explore this alternative notion of probability in this chapter.

2.1 Elements of a Propensity Interpretation

The main competitor to relative frequency interpretations is the *propensity* interpretation. This interpretation was first explicitly introduced by Karl Popper, who gave it its name (Popper 1959). However, there are many varieties of this interpretation: more, indeed, than writers on propensity interpretations. Thus, any characterization will be incomplete. Nonetheless, there do seem to be some core tenants of the interpretation. First, propensity interpretations are purported to be 'objective', and not purely epistemic, at least to the extent that such a distinction is possible. Second, they do not take relative frequency approaches to provide a (completely) acceptable interpretation of probability. Third, they take probability to be a disposition. Finally, they hold that there are probabilities of an event happening only once, that is, single-case probabilities.

2.1.1 Probability as a disposition

Prokop is unique: his full description places him in a reference class of his own—all of his own. He also thinks that there is a chance of his getting into an automobile accident, and that this chance is objective, in the sense that, given all relevant information, everyone would agree as to how big or small the chance is. And it's not the same as that of the group the insurance agent has lumped him with.

Prokop thinks the probability of his getting into an accident is a fact about him, a feature of the world, dependent on other features of the world.

For example, the chance of his wrecking his car depends on the condition of the road, his mood, the weather, and a host of other factors influencing the safety of his journey. He does not think this chance is a fact about what class of people he belongs to. In line with this he takes the view that even if he never actually gets into an accident there is still the objective chance that he could. And while this chance might be, or probably is, different for every journey he takes, and so might be very difficult to pin down in the case of individual trips, it is still an objective chance.

Prokop's intuitions provide us with the core of the propensity view. According to propensity theorists, probability is to be associated with those features leading to the occurrence of an event. The probability is only indirectly, if at all, associated with the repetition of an event. In particular, the features of the event lead to a disposition for the event or a group of events to come out a certain way: for a coin to land, say, heads (on a single toss or in a series of tosses). That is, there is some set of conditions which lead to a disposition for a particular outcome to occur. A standard way to phrase this is that probability is a relation between generating conditions and outcomes, not a relation of infinitely long sequences of outcomes. Probability is a property of an experimental set-up, and so is objective.

Probability is a disposition under this interpretation, in that it is about outcomes that occur given certain circumstances. Moreover, it's about outcomes having that disposition to occur, even if the circumstances themselves never occur. If the coin in my hand is to be melted down before it is tossed, it still has, according to this view, a chance of coming up heads if tossed, just as salt has a disposition to dissolve in water even if it never is placed in water.

It should be noted that the propensity is not a property of an object like a coin, but a property of an experiment, like tossing the coin. This includes all relevant facts about the coin and the environment in which it is tossed. If the Amazing Merlin (in town this week only) tosses this coin, its outcome will be different than if Jarda tosses it. For the Amazing Merlin, either by astonishing command of the local magnetic field or by sheer cheating, can make the coin come up however he wants it to. The chance of the coin coming up heads or tails, then, is not determined by the coin alone, but by the environment in which the coin is tossed. The probability is not determined, in this interpretation, by the relative frequencies of outcomes.

2.1.2 *Single-case probabilities*

Proponents of the propensity approach hold that the coin in my pocket has a chance of coming up heads on the next toss, and accordingly that chance is a probability (a disposition to come up heads or tails). As we shall see in the next section, this case is in fact problematic, and so we will instead stick with another paradigmatic case: radioactive decay. One main motivation for believing that there are single-case probabilities comes from cases where there does seem to be an objective chance of something happening, but no meaningful notion of repetition. Radioactive decay is the poster child for this case.

Radioactive atoms have an unstable play of forces in their nuclei. They thus have a tendency, a propensity, to break apart (usually referred to as 'decay'), emitting radiation (particles) and leaving behind other, more stable, elements. Radioactive isotopes are characterized by their half-lives: the amount of time it takes for half of some amount of the isotope to decay. This leads naturally to a probability, seemingly for individual atoms: if the half-life of an isotope is, say, 1 day, then the probability of *this* atom decaying before the end of the day is .5.

David Lewis 1994 also uses an example of radioactive decay to show that it seems quite intuitive to think that there are single-case objective probabilities. Suppose that there is a radioactive element, Unobtanium^{346}, so difficult to produce that only a few atoms of it will ever be produced during the entire life of the universe: surely, it has a half-life, and so an associated probability of decay. This decay is not a mass phenomenon, unlike other, more familiar instances of radioactive decay (for example the virtually limitless atoms of uranium isotopes). Therefore it seems quite natural to say that each atom has a propensity, a chance of decaying, at any particular time, and to add this chance to the list of its other properties like electric charge and mass. This contrasts with the relative frequency view that probability is a property of collectives, not of individuals: the chance of decay is an individual chance.

We now have the ingredients of a propensity view that matches Prokop's: probabilities are single-case objectively determinable dispositions—or chances—of an outcome occurring in a given setting. The propensity interpretation thus seems very intuitive: it would be nice to save this intuitiveness. This, as we shall now see, turns out to be a very difficult task.

2.2 Problems with Propensity Interpretations

The propensity account faces severe difficulties, some of which we shall catalogue in the following sections. First there is an ontological problem that can be phrased as a dilemma: either single-case probabilities give rise to a version of the reference class problem, or they commit us to indeterminism. The usual response to the dilemma is to embrace indeterminism and a 'universal reference class': but this in turn gives rise to a severe epistemological problem—the determination of the value of propensities seems in principle impossible. It also commits a propensity view to indeterminism. Second, the notion that propensities are a measure of causal ability and obey the standard probability calculus is subject to paradox, or contradiction, depending on your view. Third, there seems to be no reason propensities should be probabilities: but then they won't do as an explication of probabilities in physics. Finally, the one interpretation of single-case dispositional probabilities that offers some hope of avoiding these problems doesn't in fact seem to yield single-case probabilities, or, indeed, be any different than some relative frequency interpretations.

2.2.1 Indeterminism and the reference class

Howson and Urbach 1993 make the following argument: suppose we have a coin that will be tossed, say, forever, and that the relative frequency of heads will be equal to, say, .5. Further suppose that we can predict with certainty the outcome of a toss given certain parameters.[1] What, then, is the propensity of the coin to come up heads—.5 or 1? This is, as Howson and Urbach point out, the reference class problem, but in a much worse form than usual: if the conditions determine propensities, then we will have to pick out those conditions to determine relevant probabilities—that is, we either have to choose the full set of conditions, where the probability is 1, or the set of conditions that give us the relative frequency, and probability, of .5.

One response to this difficulty is to follow Miller and 'eliminate the reference class' (1994: 182). For Miller this means that the reference class is the complete state of the experiment of the universe at the time: 'Strictly, every propensity (absolute or conditional) must be referred to the complete

[1] In fact, there *is* an Amazing Merlin: Diaconis, Holmes, and Montgomery 2007 detail the building of a coin-tossing machine. They survey analyses of the physics of coin tossing, and conclude that 'coin tossing is "physics" not "random"' (2007: 211). Hacking 2008: 25 claims that you can learn to toss a coin so that it almost always lands on the side opposite to the side up when it was tossed. (I was alerted to this by Peter Milne.) Nonetheless, the Amazing Merlin can be thwarted if the coin is not caught, but allowed to bounce on the floor: bouncing makes it much more difficult, if not impossible, to compute a possible outcome.

situation of the universe (or the light-cone) at the time. Propensities depend on the situation today, not on other situations, however similar' (Miller 1994: 185–6). So, no contradiction can arise from a variation of the reference class. This, however, leads to two further problems.

The first, rather obvious, problem is that we can never know the complete state of the universe, and so the theory is completely without empirical content (Gillies 2000: 126–9—Gillies's discussion informs much of what we will cover in this section). We will return to this in the next section. There is also a second problem, related to the problem of empirical content: that of determinism. For if determinism is true and we take the reference class to be the complete state of the universe, the propensity of any outcome must be 0 or 1, for the state of the universe will determine the outcome. This propensity approach therefore must prejudge the question of determinism in favour of indeterminism. Some see this as a bug (Howson and Urbach 1993: 341), others as a feature (Giere 1976: 344). Still, if determinism is true, then the propensity interpretation will not do as an interpretation of probabilities in the physical sciences.

If we do not, however, take into account the entire universe to avoid concluding in favour of determinism, then the reference class problem arises. We could, with Fetzer (in chapter 3 of Fetzer 1981), make a requirement of maximal specificity: the reference class of the propensity, the generating conditions, as it were, are a maximal set of the laws and conditions sufficient for the determination of an event's propensity. So, for example, for the toss of a coin, this set might not include the colour of my daughter's favourite coat. This will not, of course, solve the problem, since all the relevant causes will completely determine the outcome of the coin toss in a deterministic world.

Much more can be said about the determination of the reference class—see for example Gillies 2000: 121–3. Gillies makes the point that even if we can determine a maximal reference class, it need not be unique—that is, there can be more than one maximal reference class.

2.2.2 *Empirical content*

The idea of a limited but maximal reference class might seem to address the problem of empirical content. At least we don't (necessarily) have to know the entire state of the universe, but only the perhaps manageable part which might affect the coin toss. However, we would have to determine what the relevant laws and conditions are. How we might go about this without a prior theory of objective probability is not clear, since it seems this is exactly the theory we would need to use to determine probabilities.

Another way of approaching the problem would be in terms of explanation. If we take the reference class as the entire universe, then explanations of particular probabilistic occurrences become fairly vacuous. 'Why did that happen?' 'Because that's how the universe is'. But if we try to increase explanatory context by isolating those bits of the universe that led to the occurrence, we are again in need of an explanation of why those bits figure in the explanation. But that explanation will be a probabilistic one, requiring a theory of probability, which the propensity interpretation aims to provide.

To sum up, there are two problems: an epistemological one and an ontological one. There is a reference class problem with propensities that can be avoided by taking a universal or maximal reference class. But this means that if the universe is deterministic, then there are no non-trivial probabilities. To avoid this undesirable outcome, a propensity theory must assume that the universe is indeterministic. But there is also an epistemological problem, namely, the empirical burden placed by the universal or maximal reference class. To determine probabilities is to determine the state of the universe. If we attempt to reduce the epistemological problem by taking a smaller piece of the universe, we must specify *which* piece, thus reintroducing the reference class problem. These problems can be seen as forming a dilemma: if we choose to have no reference class, then we are saddled with an unsupportable epistemological burden; but if we choose not to have a universal reference class, then we need a solution to the reference class problem. Hájek 2007 argues that a version of the reference class problem in fact plagues all the interpretations of probability discussed in this book. One version of this for the subjective interpretation can be seen in section 3.8.2. (Hájek also makes a distinction between a metaphysical and an epistemological reference class problem, but, ontologically speaking at least, to a different end.) It is also worth repeating that relative frequency interpretations face a similar problem with vacuous explanations, which is another way to frame the discussion in 1.2.5.3.

There is, as well, another class of problems having to do with the link between the probability calculus and the notion of propensities. The next section covers the charge that the two are not linked at all.

2.2.3 Humphreys's paradox

Propensities are dispositions: this is the starting point of the interpretation. There are conditional and unconditional dispositions. An example of an unconditional disposition is that of a radioactive atom's disposition to decay —this is unaffected by any known processes. But there are also conditional dispositions: salt, being put in water, is disposed to dissolve. If the weather is hot, Prokop is disposed to drink beer. Dispositions are closely linked to causes: if certain background conditions hold, then a disposition is realized: salt will dissolve in water, the light will come on, Harold will give up the ghost. We could say that the salt's dissolution was caused by the presence or absence of certain conditions (being in water with an appropriate salinity and temperature, etc.). Thus some take propensities to be 'some sort of weak causal disposition' (Giere 1976: 321–2).

This leads naturally to representing chance as a disposition for something to occur given certain conditions, i.e. we want a conditional propensity. Recall that the whole point of the propensity interpretation is to provide a foundation for the use of probabilities in the sciences. Therefore it is natural to represent conditional propensities as conditional probabilities. This leads to an interesting problem: suppose that there really is (and there really is) a propensity for Prokop to drink beer when it's hot. There is therefore a probability of his drinking beer when it's hot. But we can then derive the probability of it being hot when Prokop drinks beer from the probability of Prokop's drinking beer when it's hot. If we take the propensity account of probability representing a causal link, it looks like our drinking beer is somehow the cause, or contributes to the disposition, for it to be hot. We do not yet have a contradiction, but rather 'merely' a paradox, since we have an unexpected and, for proponents of propensity interpretations, unpleasant, result; namely, that if Prokop's beer drinking is influenced by the weather, his drinking likewise influences the weather. This is the gist of one version of Humphreys's paradox, first introduced in Humphreys 1985.

But let's discuss the paradox using some actual values. Suppose we happen to know the unconditional propensities for beer drinking and for hot weather (or that we can calculate these values by using the theorem of total probability; see Appendix A.2.5). Take

$$p(A|B)$$

to be the propensity for A to happen given that B happens. Suppose then that

$$p(\text{drink beer} \mid \text{hot weather}) = q, q > .5.$$

That is, hot weather increases Prokop's beer-drinking propensity. ('Drink beer' is short for 'Prokop drinks beer at 16:00 on 15 August 2010', similarly for 'hot weather'.)

We can now calculate

$$p(\text{hot weather} \mid \text{drink beer})$$

by using Bayes's theorem (appendix A.2.5),

$$p(\text{hot weather} \mid \text{drink beer}) = \frac{p(\text{drink beer} \mid \text{hot weather})p(\text{hot weather})}{p(\text{drink beer})}$$

Suppose we could determine the following values using the General Theory of Beer Drinking and Weather:

$$p(\text{drink beer} \mid \text{hot weather}) = .95$$
$$p(\text{hot weather}) = .6$$
$$p(\text{drink beer}) = .7$$

so,

$$p(\text{hot weather} \mid \text{drink beer}) = .95 \times .6/.7 = .81$$

Drinking beer, therefore, seems to have a strong causal influence on the weather.

Perhaps to avoid these results we could stipulate what seems obvious: that hot weather does not occur because of beer drinking. We might try to represent this probabilistically. That is, we could assume that

$$p(\text{hot weather} \mid \text{drink beer}) = p(\text{hot weather} \mid \text{no drink beer}) = p(\text{hot weather}).$$

Each of these equalities implies the other: to see this use the definition of conditional probability, and then the fact that beer drinking and hot weather are independent; that is, use the fact that if $p(A \mid B) = p(A)$, then $p(B \mid A) = p(B)$.

However, adding in this independence principle leads to contradiction given the probability assignments, since

p(hot weather | drink beer)

will be both .81 and .6 (by the independence principle and by the initial assignment of probabilities).

The source of the contradiction is clear: probabilities are reversible in a way that causal influences are not (or at least are generally agreed not to be, at least not always). So if we attempt to restrict conditional probabilities by defining them in one way, $p(A|B)$, then they must be defined in the opposite way as well, e.g. $p(B|A)$. But causes are temporally ordered. Hence, one obvious way to avoid contradiction suggests itself, namely that we tense the events over which we define the probabilities.

There is no technical barrier to introducing tenses: we need merely subscript the events: there is hot weather at time t_0, and this leads Prokop to drink beer at a later time t_1.

p(drink beer at t_1 | hot weather at t_0) = .95

But this won't help, because we can still calculate

p(hot weather t_0 | drink beer at t_1) = .95 × .6/.7 = .81

which, if propensities are taken to be causal, seems to give us reverse causation in a particularly clear way. It is also clear that any conditional propensity will involve us in reverse causation. This is, of course, problematic if we wish to represent propensities (of this causal sort) by the probability calculus. Indeed, Humphreys took his argument to show that the probability calculus cannot be used to represent propensities; most of the literature takes the opposite view, that his paradox shows that the propensity view is fatally flawed.

In the same vein as the forgoing, Peter Milne 1986 offers a particularly simple, and devastating, counterexample to taking conditional probabilities as measures of causal influence. Assume we have a fair die, and consider

p(six | the outcome of the roll was even)

Milne points out that the probability cannot be, according to such a propensity interpretation that takes probabilities to be causal, 1/3. For if the roll is even, then the outcome is either two, four, or six. But the first two outcomes are incompatible with six, making them impossible, and so of probability 0, while, obviously, the last makes the conditional probability 1. So if there is to be a causal link, the conditional probability seems impossible to interpret, since the generating conditions make the outcome either determinate or impossible. And in any case, it seems odd indeed to assert that the roll coming up even has a *causal* influence on the roll coming six. To reiterate: the propensity interpretation is supposed to give an account of probabilities as single-case chances, which leads to a notion of conditional probabilities as a causal (or causal-like) link between chances. But Milne provides a perfectly mathematically respectable conditional probability (indeed, an example of the sort the probability calculus was invented to solve) which is not causal in any way.

Humphreys's paradox has given rise to an extensive literature. Surveying it shares the difficulties of surveying the propensity approach, but luckily we have a fine survey in Humphreys 2004, to which I refer the interested reader. Still, there is one possible response to the paradox that is worth mentioning, since it seems to contain either a way to fix the propensity interpretation, or to show that the interpretation cannot be fixed at all.

Humphreys's paradox arises from a tensed reading of the probability calculus. (If the cause, the condition, occurs first, then the effect occurs with a certain probability, or has a certain propensity to occur, afterwards.) We could, however, take the conditional probabilities of the propensity approach to be untensed, in the sense that they all refer to current events: that is, the domain of the probability function are all events right now, that have a propensity to lead to certain developments in the future, described from our position right now, in the present. Thus the probability of hot weather given beer drinking represents how the two go together right now, not how one has caused the other. This probability can then be used to make predictions about the future. Humphreys attributes this response to Miller 1994 and McCurdy 1996: he terms it the *co-production* view.

Another way of putting the co-production view is that conditional probabilities do not directly represent the influence of one propensity on another. That means, as Humphreys points out, that conditional probabilities are not conditional propensities, if conditional propensities are to be taken as the influence of the conditioning event. Humphreys takes this as an argument against this interpretation of propensities:

> A major appeal of single-case propensities has always been their shift in emphasis from the outcomes of trials to the physical dispositions that produce those outcomes. To represent a conditional propensity as a function of two absolute propensities, as co-production interpretations do, is to deny that the disposition inherent in the propensity can be physically affected by a conditioning factor. This is, at root, to commit oneself to the position that there are conditional probabilities but only absolute propensities.
>
> (Humphreys 2004: 275)

However, denying the causal character of propensities will avoid the paradox, and might remain attractive for some. Still, Peter Milne has raised the following problem in correspondence. If we accept that propensities of future events are to be understood as propensities *now*, not in the future, how are we to understand those propensities once future events come to pass? It would be natural to assume that the propensities change in accordance with conditionalization—that the conditional propensity of the future event becomes the absolute propensity of that event at that time. But this reintroduces the notion that propensities are not future propensities now, but also propensities in the future. This in turn reintroduces the notion of propensities changing over time, which brings back Humphreys-type problems, since the inverse conditional probability may now be interpreted as going backwards in time.

2.2.4 Why are propensities probabilities?

If we approach propensities as *sui generis* components of the world, arising from certain physical situations with certain associated causes, as many propensity theorists do, it's very hard to see why they should be probabilities at all. In fact, some authors (notably, Fetzer 1981) argue that they are not. But we would then lack an interpretation of the probabilities that are actually used in science like the probability of radioactive decay.[2] This would mean that the propensity interpretation would fail to be a complete explication of probability in the physical sciences. But this is where we find the best candidates for objective probabilities as propensities. And we would also be left wanting an explanation of why these entities, propensities, have been introduced into our metaphysical discourse.

One response is that we take propensities to be probabilities because this is the simplest way of dealing with probabilities (Mellor 2005: 57). However, this assumes that the simplest explanation in this case is the best explication—an argument for which Mellor does not supply. Secondly, Mellor's response turns on what is meant by 'simplest'. There are infinitely many functions we could choose to represent propensities: why use probabilities? Perhaps what Mellor means is that there are many functions that

[2] This is not to say that all probabilities in science need satisfy the Kolmogorov axioms. Quantum mechanics, for example, may require a different axiom system, as may even statistical mechanics. It is a question whether a version of Humphreys's argument would apply to these systems. (My thanks to an anonymous referee for pointing this out.)

can be scaled to a probability function: and, indeed, many find the probability function to be the easiest to work with. This is true (and is of central importance in Cox's argument for subjective probability in section 3.7). But the question then arises: why pick a function that can be scaled to a probability? There are many other easy-to-work-with functions that cannot be scaled to probabilities. Therefore, some argument must be given for using probabilities as opposed to these other functions.

2.2.5 *Are propensities relative frequencies?*

The question of the relation between probabilities and propensities leaves unanswered, however, the relation between propensities and relative frequencies. For example, it seems quite unlikely that the exact conditions under which an experiment takes place can ever be repeated, if we require that 'exact conditions' means the state of the universe. Indeed, it seems rather unlikely even given the weaker reading of 'all relevant causal circumstances'. Hence, to get the notion of repeated experiments, and so the notion of relative frequencies, we must find some way to relax the requirement of 'exact conditions'. How to relax the requirement in a purely objective framework is, of course, the problem of the reference class.

Suppose, however, that we have solved the problem of what counts as a repetition of an experiment. Then perhaps we will get non-trivial relative frequencies, i.e. between 0 and 1. This would show how to get relative frequencies from propensities. But then, again, we may have no relative frequencies that converge to a particular value: perhaps they will just oscillate around different values forever. Von Mises postulated convergence—but nature may not grant us convergence. Hence, propensities may not be related to relative frequencies. This seems quite unwelcome as a conclusion if we are aiming to give an interpretation of the probabilities used by the sciences.

Mellor 1971, chapter 4, suggested an account of propensities as dispositions that solves this problem (and a number of other problems). He argues that propensities are not some kind of faulty disposition that sometimes are realized given the occurrence of the proper conditions, and sometimes not. That is, Mellor distinguishes between *tendencies* and propensities. A tendency is a faulty disposition: sometimes, but not always, given the appropriate conditions, the disposition will be actualized: the coin has a tendency to land heads, sometimes, given the appropriate conditions. This does capture one notion of propensities, that they are dispositions which are sometimes realized given the conditions, and the strength of the tendency is taken to be the probability. But, as Mellor points out, the notion of a tendency is in need of explanation itself, and it's hard to see how it could be explained non-circularly, for it seems that we would have to explain tendencies by reference to the notion of a propensity.

Mellor instead chooses to take propensities as full dispositions, but dispositions of a special type: their realization, known as their display, is a *probability distribution*. So at each repetition of an experiment, we get the display of the disposition, but not always the same result. The repetition gives us, over time, a distribution of likely results. Propensities are then dispositions of trials to produce certain relative frequencies. Naturally, we will need an explanation of propensities as well, but they will be explained as dispositions, which are familiar entities in the sciences.

As an example, consider the roll of a fair die. On each roll we get a result: 1, 2, 3, 4, 5, or 6. But the distribution is that the relative frequency of each outcome is 1/6. And so over time, the distribution is realized through the repetition of the roll, and we get a probability of 1/6 for each of the outcomes: we get a flat distribution (pictured in section 4.1). But an unfair die will have a different display: its distribution will be skewed, that is, some outcomes will be more frequent than others.

This account gives a convincing link between relative frequencies and dispositions. But it does have one serious flaw in terms of a propensity account: as it stands, it is not a single-case objective propensity interpretation. The probability distribution may be realized at each repetition of the experiment, but that does not, in any meaningful sense, tell us the chance that the next roll of the die will be a 3. It only tells you that in a long series of runs, 3's occur 1/6 of the time. (We will see how Mellor and others approach this problem outside of the framework of objective probability in 4.2 and 4.3.)

2.2.6 Is there a separate propensity interpretation?

It is now time to consider the differences between von Mises's relative frequency interpretation and (some version of) the propensity interpretation. For it turns out that von Mises did agree that some properties of an experiment give rise to a collective—that is, he makes explicit reference to the generating conditions on at least one occasion. He gave the example of two pairs of dice, one of which is biased (having been 'loaded'—tampered with). In fact, with the biased pair, 'at least one 6 appears at nearly every throw'. He then continues each pair has a characteristic probability of showing 'double 6', but these probabilities differ widely.

Here we have the 'primary phenomenon' (*Urphänomen*) of the theory of probability in its simplest form. The probability of a 6 is a physical property of a given die and is a property analogous to its mass, specific heat, or electrical resistance. Similarly, for a given *pair of dice* (including of course the total setup) the probability of a 'double 6' is a characteristic property, a physical constant belonging to the experiment as a whole and comparable with all its other physical properties. The theory of probability is only concerned with relations existing between physical quantities of this kind.

(Von Mises 1957: 13–14)[3]

Von Mises's approach could be seen as reducing the theoretical term of probability to the observation of frequencies, and hence, as operationalist. Operationalism may be thought of as the view that the meaning of concepts in science is tied to the way in which they can be measured. Gillies (1973: 37–42 and 2000: 100–1) documents von Mises's operationalist and positivist leanings. Of course, there is not much doubt that von Mises was a positivist, having written a textbook on the subject (his 1939 *Positivism*). Gillies argues that von Mises's operationalism stops him from being able to refer to generating conditions (*qua* dispositions). Gillies further argues that this means that von Mises's cannot meaningfully assign probabilities to trials actually carried out, as the propensity theorist can.

Yet, there seems to be no barrier to jettisoning any unnecessary operationalist luggage, and adopting propensity metaphysics, when necessary, to explain where collectives come from. Indeed, it is quite difficult to see how a theory could really refer only to frequencies alone: we must pay *some* attention to generating conditions to separate some collectives from others, since the collective generated by flipping a coin on the moon is different than that of flipping a coin on the Earth outside on a windy day in the presence of powerful fluctuating magnetic fields. (This is a modification of an example from Popper 1959, who felt that considerations of this sort necessitated a propensity approach—I think he was only half-right, but see Gillies 2000: 115–16 and Howson and Urbach 1993: 339–40.) Therefore, Mellor's account of the display interpretation fits quite nicely with von Mises's account: we get a version of single-case probabilities to account for decay, but still have the link with relative frequencies.

[3] Howson and Urbach 1993: 338, point to this quote. Von Mises unfortunately gives two kinds of propensity interpretations here: in the first, where he considers the single die, the propensity is seen as inhering in the die; in the second, the propensity is part of 'the experiment as a whole'. So I shall ignore the first as a slip.

Tying von Mises's view to a propensity view makes it vulnerable to the objection, discussed in section 2.2.1, that probabilities would only be non-trivial in a deterministic world, thereby committing von Mises to an indeterministic view of the world. But his view is *already* vulnerable to such an objection: for if we know how a sequence of members of a collective is generated, then, in a determinist world, we could change the relative frequencies of subsequences by appropriate choices. In fact, since we could determine (given God-like-enough powers) when an outcome would occur, or not, we could create subsequences with any probability we chose. And, anyway, von Mises explicitly ties himself to an indeterminist view of the world in the final sections of *Probability, Statistics and Truth* (1957).

But, if we adopt this view of propensities, the interpretation no longer seems to be interestingly different from a relative frequency interpretation, albeit a more metaphysically subtle relative frequency interpretation. In particular, the notion of propensities as a sort of weak causation is lost. This may or may not be an objection, depending on how attached one is to having a different propensity interpretation, or having an interpretation of propensities as causal.

2.3 Conclusion

There are a host of other charges aimed at these propensity interpretations which I have not addressed (but which are discussed in, for example, Anthony Eagle's 2004 'Twenty-one Arguments against Propensity Analyses of Probability', as well as in Gillies 2000). While I think that a version of the propensity account could be constructed to address these difficulties, it is clear that many other versions of the propensity interpretation do not avoid them.

As I said at the start, there is a great wealth of propensity views. I have not begun to cover the totality of these views. I offer a rough and ready classification based on the taxonomy found in Gillies 2000 and enriched by Mellor 2005 to show how much I have left out. Propensities can be:

(1) degrees of possibility
(2) partial dispositions related to the outcome of an experiment/chance set-up which
 (a) give rise to relative frequencies in a possibly infinite sequence of outcomes

(b) give rise to relative frequencies in a long but not infinite sequence of outcomes as a relative frequency
(c) may or may not have anything to do with frequencies, being simply dispositions to occur in one particular instance

(3) full dispositions of a chance set-up which give rise to a probability distribution, which then determines relative frequencies.

These must be further enriched by an account of *conditional* propensities in order to account for the difficulties raised by Humphreys's paradox (a classification of which can be found in Humphreys 2004).

I have not discussed (1), for example. One reason is that it will, in a sense, be covered in Chapters 5 and 6. For another, it faces technical problems that seem to severely limit its use as an explication of probabilities in science (but see Mellor 2005). I have briefly mentioned (2) in the discussion of Mellor's account of propensities as full distributions (section 2.2.5), but not the sub-options. Instead, I concentrated on (3), which seems to give the most promising account of propensities.

Yet, even if a successful propensity interpretation is developed, it will require supplementation with an account of induction. As mentioned already, the interpretation of objective probability need not tell us how we determine what those probabilities are. The latter is a part of epistemology; the former need not be. This is particularly clear in von Mises's interpretation —he adopted a version of the theory outlined in the next chapter, while, of course, advocating a relative frequency interpretation of probabilities in the sciences.

At this point we are faced with a forking of paths. There are a number of accounts of probabilities in specific sciences: biology and physics are the main examples. Probability finds a natural home in statistical mechanics and in quantum mechanics, for it is at the core of the theories. Probabilities also permeate biology. There are examples from biology in most introductory textbooks to the mathematics of probability, very often from genetics.[4] We will not pursue these interpretations, because our aim is simpler—to lay out as generally as possible some standard interpretations of objective probability, and note that they still leave room for, and indeed require, other

[4] For the curious, Sklar 1993 is a good introduction to philosophical questions about probability in statistical mechanics. Strevens 2003 is a recent noteworthy contribution. The study of probabilities in quantum mechanics is a complicated field: a good entry is chapter 6 of van Fraassen 1980—I've not actually found much else. Sober 2008 and Rosenberg and McShea 2008 are good starting points for the notion of probability in biology.

interpretations of probability. The hard work of testing their adequacy in particular sciences is a different task.

This leads to another fork. There are accounts of probability that purport to be objective and also epistemic. These come from a tradition of methodological falsificationism, which takes the aim of science to be that of falsifying statements, not of confirming them. Standard accounts of this approach may be found in any introductory statistics textbook. This is not a debate we will canvass, since while closely related, it is, with the exception of Gillies's view to be discussed in the next paragraph, not directly about the interpretation of the probability calculus. The reader may consider this a flaw if they wish, and consult, for example, Howson and Urbach 1993 for one view, Mayo 1996 for the opposite.

Donald Gillies (1973 and 2000) develops an interpretation that he identifies (in 2000, but not 1973) as a propensity interpretation, in that it is a dispositional account of probability. It does not, however, yield single-case probabilities. The aim of the theory is to provide an account of probability relative to long, but not infinite, sequences of outcomes, compatible with aspects of Karl Popper's falsificationism. In turn, Gillies argues that his theory gives empirical significance to probability statements (while avoiding the subjectivism inherent in the approaches to be discussed in the next two chapters). In particular, his Falsifying Rule for Probability Statements serves to introduce the notion of probability as a theoretical entity, the value of which can then be checked by significance testing. Howson and Urbach (1993: 335–6) criticize Gillies's theory; Gillies responds to these criticisms in 1990. We won't pursue this debate because it does take a very different methodological approach: one of falsification as opposed to confirmation. Again, the reader may consider this a bug, or a feature.

So, we take it that there is still room for something else to be said concerning the notion of objective probabilities: this something else will occupy the remainder of the book. The notion we will entertain is an epistemic interpretation, that of probabilities as degrees of belief.

3

Subjective Probability

3.1 Introduction

Prokop goes to pick up his friend Vladimír from the airport, who has flown in from Prague for a visit. ('Cool shoes, man'. 'Thanks, they were on sale'.) After a hair-raising ride in heavy traffic, which Prokop makes even more frightening by reciting statistics about car crashes and the American health care system, they arrive at Prokop's flat. Later that evening, they drink home-brewed beer ('Homesick Pilsner') and watch reruns of Star Trek. During 'Errand of Mercy' Kirk asks 'What would you say are the odds on our getting out of here?' Spock replies 'Difficult to be precise Captain. I should say approximately 7824.7 to 1'. Prokop and Vladimír laugh loudly, disturbing the neighbours.

Later, after a serious dent has been put in the beer supply, the conversation turns to Margaret Thatcher. Vladimír is sure that Mrs Thatcher resigned sometime after 1995, Prokop is sure that she resigned before. They decide to bet: if Mrs Thatcher resigned before 1995, Vladimír will buy Prokop a bottle of Fernet. If she resigned after 1995, Prokop will buy Vladimír a bottle. Fernet is much more expensive in America than in the Czech Republic.

Prokop's studies in physics are going quite well. He regularly watches the weather forecasts as a way of procrastinating, but it doesn't work, since he starts to think about the physics of weather systems. In America weather forecasts give a probability of precipitation on the next day ('Tomorrow will be partly cloudy, with a 25 per cent chance of rain'). Researching this in the library, Prokop learns that weather forecasters are using a methodology for checking their own forecasts that resembles betting.

Betting seems to Prokop to be a large part of American life. Even though recently outlawed, Internet betting is very popular. And so-called prediction markets (for example, the Iowa Electronic Markets), he notes, are nothing other than people getting together to bet on what will or will not happen. There is a lively trade on an option on whether the next president will be a Republican. In his research he finds that bets can be represented as probabilities, and these probabilities are not necessarily of mass phenomena. The interpretation commonly used as foundations for the examples provided is one that takes probability to be a representation of a degree of belief. It is known as the Bayesian, subjectivist, or personalist interpretation.

There are a number of arguments in favour of the Bayesian interpretation. In the following sections, we will cover a number of them. We will also examine some successes claimed for the interpretation, as well as reasons to doubt that they really are successes.

3.2 Dutch Book Arguments

Let us return to Prokop and Vladimír. Abbreviating 'The event happened', or 'True' as T, and F for the opposite, the following table represents the terms of the bet from Prokop's perspective:

Mrs Thatcher resigned before 1995	Pay-off
T	+bottle of Fernet
F	–bottle of Fernet

If Mrs. Thatcher did resign before 1995, Prokop gains a bottle of Fernet, and if not he buys Vladimír a bottle. (Bets will be taken to be about propositions—it is difficult to see how you could bet on something that couldn't be described by a proposition.) This basic scenario can be the basis for developing a much finer picture of the strength of belief. Suppose that Prokop is very sure indeed of the date of Mrs Thatcher's resignation. He might offer the following bet: If I'm right, you give me a bottle of Fernet, and if I'm wrong, I'll give you two bottles of Fernet. Prokop would be 'putting his money where his mouth is'.

But there's a limit to how much we can associate certainty with willingness to lose bottles of Fernet. They're heavy, moderately expensive, and you can only drink so much Fernet. We could use the small bottles of Fernet, or just individual drinks of Fernet. But, again, you can only drink so much. And not everyone drinks alcohol, or even Fernet. Fernet, then, whatever its other merits might be, is not an ideal currency for betting. So why not just use the currency we already know?

A much better currency for bets would be money—not a lot of money, but, say, quantities of cents and euros or dollars. If you are *really* sure about some proposition you should be willing to risk more relative to potential gain. Say, you might want to put up 50 cents, which you will lose if you're wrong, while only requiring your betting partner to put up 10 cents, to be paid to you if you are right. The ratio of potential gain to potential loss is known as the *odds*. In our example, the odds are 5:1. (Of course, we could use anything of value that we could denominate in a relatively simple fashion, but money is traditionally used for the purpose of betting.)

The practice of betting I describe is somewhat different than the one the reader is probably used to. For one thing, betting shops take a portion of your winnings, so odds do not perfectly match what you gain when you win a bet. For another, informal bets made, for example, in the pub, often are of the 1:1 variety (if I'm right you get a bottle of Fernet, if not I get a bottle), where the uncertainty is captured (in a crude way) by the currency: the more expensive the item on offer, the greater the certainty in the proposition (or its negation) under consideration. Also, some seem to regard betting as a game of pure chance, and seek to engage in bets on topics of mutual complete ignorance (say, betting on the outcome of a coin toss, or on the identity of a figure half-glimpsed in bad light). These bets are too crude for our purposes.

Given fine-grained enough currency, we have the makings of a device to measure degrees of belief. The more strongly you believe something, then, if you have enough money, and are relaxed about betting, the better odds you will offer to your opponent. For example, I'm very sure that the Vltava flows through Prague. So, if anyone offers to bet with me, I'll give very good odds—say 1000 to 1. On the other hand, I neither know, nor, really, care, whether it will rain tomorrow in, say, Dusseldorf (at least, right now I don't care: this can change). So I would give you even odds, if I were inclined to bet. Similarly, as I write this, I don't think it's not going to rain in the next two hours, and so I'll give you low odds on a bet for dry weather. (The double negation tells us something about how bets function, as we'll see.)

3.2.1 Fair bets

Before we can develop the main thesis of this chapter, we will need to introduce the notion of *fair* betting odds. To illustrate, suppose that we are betting if tram 22 stops at Krymská on its normal route. I am quite certain of this, and so am willing to offer 40:1 odds in a bet that it does. But I am also willing to accept odds of 39:1, since I reduce my risk of loss to 39. I would also accept 38:1, reducing my potential loss to 38. Indeed, I would accept any bet down to 0:1.

But suppose that I had to bet that the tram does not stop at Krymská: that I had to bet *against* the proposition that the tram does stop at Krymská. Since I in fact do believe that the tram stops at Krymská, I am only willing to offer odds like 1:40 (if false I pay out 1, if true I get 40). I would similarly accept any number above 40 on the right-hand side (and any fraction of 1 on the left-hand side).

So how do we express the maximum amount I am prepared to risk for the potential gain? The usual device is to imagine that the person you bet against can change the direction of the bet. That means that you don't know if you are to bet on or against the proposition. This motivates you to find what you actually think the odds of the event are that will minimize your potential loss (or maximize your potential gain) to each side of the bet. In the case we just used, the odds are 40:1. (The assumption that you think that there are odds, much less correct odds, associated with the event, carries with it a fair amount of philosophical baggage, as we shall see in following sections.)

Another way to put it is that fair odds are those odds that you do not think give either side of the bet an advantage. (That is, they are odds you think equalize potential loss and potential gain.) And fair odds should represent what you think the actual odds of the proposition being true are. Fair odds play a special role in Bayesian theory, as we shall see at the end of the next section.

Not all bets are equally good. In fact, some bets are simply stupid. Prokop says to Vladimír 'If Mrs Thatcher resigned before 1995 I get a bottle from you, and if she resigned after, you owe me a bottle'. This is like the coin-tossing game 'heads I win tails you lose': if you accept this bet, you will always lose. The technical name for such a bet is a Dutch Book, although perhaps a neutral term like 'stupid' would be preferable. (The origin of the term is mysterious. My preferred explanation is that a bet can be called a book, and 'Dutch' was a general term of abuse in English, e.g. Dutch Date, Dutch Uncle, and Dutch Courage. There is, alas, no evidence that this explanation is correct.) The Dutch Book *theorem* establishes that all and only non-stupid bets obey the axioms of the probability calculus: that no Dutch Book can be made against someone who follows the probability calculus (this is sometimes called the Ramsey-de Finetti theorem). The upshot is: to avoid stupid bets, have the bets conform to the probability calculus. In terms of fair bets, it says that all and only non-stupid bets can be fair, and so must obey the probability calculus. In the next section I will present this argument in detail. It requires a bit of algebra, and so the tired reader may skip these bits if he or she is willing to accept my claim that the theorem, given certain assumptions, holds.

3.2.2 The forms of bets

What follows has become traditional. The clearest presentations in the literature are those of Skyrms 1986 and Howson and Urbach 1993. There are, however, a number of ways to express the argument: I stay close to Howson and Urbach.

To continue: we say a bet is *on* a proposition A if the bettor gets something of value, say, a, if A is true, and loses something of value, say, b, if A is false. A bet *against* A is where the bettor gets b if A is false, and loses a if A is true. (How do we know if A is true or false? Normally this isn't a problem. Consider the Vladimír-Prokop bet: they can settle it by consulting Wikipedia. But if Vladimír won't take the Wikipedia entry as conclusive, they might have to go to the library. And if he's really being stubborn, which, by the way, he never would be, they might have to visit the Oracle, who answers all questions to everyone's satisfaction. Yes, we *are* dealing with an idealization.)

In section 3.2 we introduced a 'pay-off table'. Below is the more general form of a bet on A:

A	Pay-off
T	$+a$
F	$-b$

We have, however, with this device performed a trick: a and b now serve as the values of a and b instead of things a and b; in other words, we take value to be denominated in some sort of currency. This is not a trivial step, as we shall see in 3.4. But for now, I would beg the reader to take a and b to be some sort of currency, say, coins that can be cut up very, perhaps infinitely, finely.

Odds are, as we said in section 3.2, defined as $b{:}a$. Some (like me) find odds difficult to work with. Luckily, using some elementary algebra we can rewrite the table so that we have the information contained in odds, but have a function ranging between 0 and 1 (inclusive). The usual way to do this is to use normalized odds: we take the odds ratio, and divide it by 1 plus the odds ratio, i.e. $p = (b/a)/(1 + b/a)$. (Any positive quantity x can be put between 0 and 1 in this way, that is, for any positive x, $x/(1 + x)$ is between 0 and 1.) The normalized odds are, as simple algebra will confirm, $b/(a + b)$.

For an illustration, consider a bet on A where the bettor is quite certain of A, and offers 9 to 1 odds on A. That means that the bettor is willing to pay 9 if A is false, and get 1 if A is true. This translates to $p = .9$. The total money at stake is $a + b$ (i.e. the amount of money at stake—recall that the punter stands to gain not only a, but lose b as well). Let us call this S. We can now rewrite the pay-off table as

A	Pay-off
T	$S(1 - p)$
F	$-Sp$

p is for obvious reasons known as a betting quotient. In our arguments we want the values a and b for which p is a *fair* betting quotient, i.e. one which is believed not to be advantageous to either side of the bet. (We will show that fair betting quotients are probabilities, which is why we choose the symbol p.) Another way that it could be put is that p is the value given if the direction of the bet can be switched by the bookie (changing a to $-a$ and $-b$ to b). This means that if p is given as a fair betting quotient, then the bettor should be indifferent to betting on or against the proposition. (This explanation of betting ratios and pay-off tables, while traditional, comes from Skyrms 1986 and Howson and Urbach 1993.)

Before we discuss how not to bet, there are two technical issues that need clarification. The first is that, as we have already said, our bets are couched in propositions. This is vague, but not dangerously so: it allows us to retain the full power of the probability calculus, moving between set-theoretic notations and logical notions. We will use the usual notation of &, $\neg$, and $\vee$ for conjunction, negation, and disjunction respectively. The second issue is the question of which propositions: we will assume that the propositions we deal with are drawn from some basic set of propositions by compounding these propositions with conjunctions and negations (so, if A and B are in the set, so is $A\&B$, as are $\neg A$ and $\neg B$). This set forms a *field* (Appendix A.2.1). The probability function we will construct will be defined over this field.

3.2.3 *How not to bet*

In section 3.2.1 I mentioned silly bets. Here is one: $p(A) < 0$. For a bet on A the pay-off is always positive no matter if A is true or false, and the pay-off *against* is negative. So the bettor on always makes a gain, and the bettor against always loses. To regard such a bet as fair is to be indifferent between a sure loss and a sure gain. By definition such a bet cannot be fair: one side of the bet always has an advantage. So, if p is to be fair,

(i) For all $A, p(A) \geq 0$

A similar argument can be made for bets on tautologies, i.e. statements that are logically true. Since a tautology is necessarily true, we need only consider the pay-off $S(1 - p)$. We need not consider $p > 1$, since that's impossible. We need thus only that if $p < 1$, the bettor against will always lose money and the bettor on will always win money. But this is obvious. Thus, $p = 1$ represent the only fair odds on a tautology. Hence,

(ii) For all tautologies T, $p(\mathrm{T}) = 1$

3.2.4 *Adding up bets and probabilities*

The next argument concerns two propositions A and B which are mutually exclusive, i.e. they cannot be true together. We will show that if we bet on A and B separately, we should bet as if we were betting simultaneously on A or B, that is, on $A \vee B$. Since we are dealing with a number of bets at once, it is convenient to assume that the stakes in all bets are normalized to the same sum S in whatever currency we use (this does not lead to a loss of generality). Thus usual practice is to set S to 1, but I will proceed in the same style as before. We denote the betting quotient for A as p and for B as q. Since A and B can't happen together, we only need the following three-row pay-off table for the combined bet on A and on B:

A	B	Pay-off
T	F	$S(1-p) - Sq$
F	T	$-Sp + S(1-q)$
F	F	$-Sp - Sq$

But this is just a bet on $A \vee B$ with a betting quotient $p + q$, as can be seen by rewriting the above pay-off table setting $p + q = r$

A	B	Pay-off
T	F	$S(1 - r)$
F	T	$S(1 - r)$
F	F	$-Sr$

which is the same as

$A \vee B$	Pay-off
T	$S(1 - r)$
F	$-Sr$

But, since they are true under the same circumstances, the two separate bets on A and on B are the same as the single bet on $A \vee B$. So, it seems reasonable that the separate bets should add to the combined bet. We can prove this: suppose that someone offers betting quotients p, q, and r on A, B, and on $A \vee B$ such that $p + q \neq r$. Thus, they think these are fair odds. So, by definition, the bettor would be indifferent between betting *on* A and *on* B and *against* $A \vee B$ with those betting quotients, by definition. The pay-off table for all these three bets is just the pay-off tables of A and B minus the pay-off table for $A \vee B$, which, with some simplification, is

A	B	Pay-off
T	F	$S(1 - (p+q)) - S(1 - r)$
F	T	$S(1 - (p+q)) - S(1 - r)$
F	F	$-S(p+q) + Sr$

Rewriting we get

$A \vee B$	Pay-off
T	$S(r - (p+q))$
F	$S(r - (p+q))$

If $r > p + q$, this set of bets would lead to a sure gain, and if $r < p + q$, to a sure loss. So the bets can only be fair when $r = p + q$. Hence, if A and B are mutually exclusive, their betting quotients must add to be fair, establishing:

(iii) $p(A \vee B) = p(A) + p(B)$, if A and B are mutually exclusive

This completes the Dutch Book arguments for the axioms for unconditional probability (finitely additive probability, that is: see Appendix A.2.4).

3.2.5 *Conditional bets and probability*

Finally, there is a Dutch Book argument for the fourth axiom. To explicate it, we must introduce the notion of a conditional bet, which is a bet on a proposition A given that a proposition B comes to be accepted as true. If B does not come to be accepted as true, then the bet on A is called off. (Example: 'I'll give you 4 to 1 odds that if Jarda comes to the party, he'll give Pavel a great big sloppy kiss'. 'You're on!' But Jarda doesn't come, so the bet is, alas, called off.) There is a direct way to prove that the bets must fit together in accordance with the probability calculus to avoid sure loss, but it involves the use of linear algebra. Those interested should consult Gillies 2000: 58–65, who, however, takes a somewhat different approach to the argument. (In this section I first followed Hacking 2001: 167–8. His argument, however, makes use of a strong assumption, strict coherence, see Appendix A.6.1, as was pointed out to me by Peter Milne. The proof I use is a modification of Hacking's suggested by Milne.)

We will take a circuitous route to the fourth axiom. All we need to show to prove it is that in certain circumstances betting quotients not obeying the fourth axiom lead to a sure loss (or gain), and so cannot be considered fair. This is enough to establish that betting quotients violating the fourth axiom cannot be fair. The situation is somewhat artificial, which once again underlines the power of the use of linear algebra.

We consider the following three bets: a bet that both A and B happen together, that is, a bet on $A\&B$, with a betting quotient of q, a bet on A conditional on B with betting quotient p, and a separate bet on B with betting quotient r. The claim is that only when $p = q/r$ can the bets be fair. Recall that betting quotients which are fair imply that a bettor is willing to take either side of the bet. Therefore, you should be willing to accept bets on $A\&B$ and against $A|B$ and against B. (We are only taking $A|B$ to be shorthand for 'a bet on A, conditional on B': for another way of approaching the meaning of $A|B$, see Milne 1997.)

To make our argument work we will consider different stakes across the bets, which are determined by multiplying a betting quotient against the unit of currency. We will set the stake in the conditional bet and in the bet on $A\&B$ to be S units of currency, and the stake in the bet on B to be pS units. The betting quotients will be: on $A|B$, p, against $A\&B$, q (the same as the stake on B) and on $B r$. It might seem odd that we set the stake on B to be pS. This, however, shows us that there is a way to lose a Dutch Book for this combination of bets, and thus will give us the fourth axiom. (I do not claim that the assumption that we can vary the size of the bet on B with respect to the betting quotients on $A\&B$ and $A|B$ is trivial. We will discuss the implications of such assumptions in section 3.4.)

A	B	Pay-off of A conditional on B
T	T	$S(1-p)$
T	F	0
F	T	$-Sp$
F	F	0

The pay-off table for $A\&B$ is as follows

A	B	Pay-off on $A\&B$
T	T	$-S(1-q)$
T	F	Sq
F	T	Sq
F	F	Sq

And a bet on B is

B	Pay-off on B
T	$Sp(1-r)$
F	$-Spr$

Putting it all together we can see that the combined bets have the following pay-off table

A	B	Pay-off on all bets
T	T	$S(1-p) - S(1-q) + Sp(1-r)$
T	F	$0 + Sq - Spr$
F	T	$-Sp + Sq + Sp(1-r)$
F	F	$0 + Sq - Spr$

Simplifying,

A	B	Pay-off on all bets
T	T	$S(q-pr)$
T	F	$S(q-pr)$
F	T	$S(q-pr)$
F	F	$S(q-pr)$

The only fair bets are where $pr = q$. All the others automatically win (or lose). This gives us, as required, $p = q/r$. This establishes

(iv) $$p(A \mid B) = \frac{p(A\&B)}{p(B)}$$

This completes the argument that betting quotients must (necessarily) adhere to the axioms of probability to be fair. We have not shown that betting quotients that adhere to the probability calculus are (necessarily) fair. In other words we have shown that adherence to the probability calculus is a necessary condition for fairness.

But the question remains whether adherence to the probability calculus is enough, whether it is a sufficient condition for fairness. Perhaps, the reader might wonder, there are other requirements that are also needed to ensure that a set of betting quotients is fair. The answer is no: there are not. This is known as the converse Dutch Book theorem. This is proved by showing that in a group of bets, where S and p are as before, if p obeys the probability calculus, the expected gain or loss is 0—*expected* since it is the probability of the individual bets multiplied by the pay-offs. (The proof of the converse Dutch Book theorem can be found in Howson and Urbach 1993: 84–5.)

3.3 The Application of Subjective Probabilities

But where does all this arguing get us? So what if probabilities can be represented as fair bets? The philosophical punch of the argument is that, as we saw with Vladimír and Prokop in 3.1, under certain circumstances, bets can show how strongly someone believes something, and so can be used as a measure of belief. The longer the odds one is willing to give to the falsity of a hypothesis—the higher the probability—the more confidence, or credence, in the proposition under consideration. The shorter the odds, the lower the probability, confidence, or credence. Betting the same both ways, 1 to 1, is probability .5, and is the same as indifference between the truth and falsity of the proposition under consideration. Probability 1, betting everything on the truth of a proposition, expecting nothing in return, is total confidence in the truth of the proposition. Probability 0 is total disbelief, or total confidence in the falsity of the proposition. So, the probability calculus could be taken as representing a scale of strength of belief (beliefs with probability between 0 and 1 are sometimes called 'partial' beliefs).

The Dutch Book theorems then provide consistency requirements for beliefs, via the third axiom, which constrains the strengths of *combinations* of beliefs. This is the *Dutch Book argument*: bets represent degrees of belief; bets are consistent if and only if they obey the probability calculus; hence, our degrees of belief are consistent if and only if they obey the probability calculus. There is obviously much to fill in; the later sections will be concerned with how to do so. That beliefs should be constrained by (some version of) the probability calculus forms the basis of a class of epistemologies, the core of which we will cover in the next section.

3.3.1 Bayes's theorem and Bayesian epistemology

Bayesian epistemology interprets the constraints of probabilities as proper constraints on partial belief, and, as we shall see, conditional probabilities are proper constraints on changing beliefs in light of evidence. Proponents hold that the probability calculus is the proper logic of partial belief, and thus the logic not only of science, but of any reasoned consideration of evidence whatsoever. These claims are made both on the basis of Dutch Book (and other) arguments, as well as on the basis of Bayesian case studies of reasoning in science and elsewhere. Support for Bayesian epistemology thus comes from 'below', from the foundations, and from 'above', from the applications.

Bayesian epistemology is named after the Reverend Thomas Bayes, discoverer of a theorem that bears his name. Bayes's theorem,

$$p(h|e) = \frac{p(e|h)p(h)}{p(e)},$$

is easily proved from the axioms (to be found in Appendix A.1). Bayes's theorem is interpreted as follows:

- $p(h|e)$ is the *posterior* probability of h, the probability of a hypothesis h given some evidence e.
- $p(e|h)$ is the degree to which the hypothesis predicts the evidence.
- $p(h)$ is the *prior* probability of h, is the degree of belief in the hypothesis.
- $p(e)$ is the probability of the evidence occurring at all.

$p(e)$ is more easily understood by the theorem of total probability,

$$p(e) = p(e|h)p(h) + p(e|\neg h)p(\neg h),$$

which shows how to consider the probability of e in light of the weight assigned to it by the hypothesis. (See Appendix 2.5 for the most general version.) The quantity $p(e|h)$ is these days known as the likelihood, and is often taken as being given by the hypothesis, while the quantity $p(h)$ puts the 'subjective' into 'subjectivism', since it represents your degree of belief in the hypothesis.[1] Bayes's theorem shows how to combine the two, hence the great interest in what could be seen as a trivial equation.

Bayesianism captures a number of relations of support between hypotheses and evidence. If our accepting e leads us to believe that h is true, that is, if $p(h|e) = 1$, we say that e verifies h, and if e leads us to believe that h is false, that is, $p(h|e) = 0$, we say that e falsifies h. But of course most evidence is not of this extreme variety. More common is when e makes firmer our belief in the truth of h, that is, if $p(h|e) > p(h)$, h is *confirmed*. Similarly, if e weakens our belief in h, $p(h|e) < p(h)$, we say h is *disconfirmed*. Changing our degree of belief in h, if e is accepted as true, setting $p'(h) = p(h|e)$, is known as *conditionalizing* on e. We can illustrate these relations with some homely examples in the first half of Box 3.2. Conditionalization will come to centre stage in section 3.8.

3.3.2 Example: Beer

A more complicated example follows. In fact, a much more complicated example. I include it for two reasons. First, I like beer. Second, discussing a real world example shows just how complicated Bayesian explanations can get. This example concerns one of Prokop's hobbies, home brewing. To make beer, we need to extract fermentable sugars from a grain, in the West usually barley. The sugars are then consumed by yeast, producing alcohol as a by-product. The first stage in this process, excepting the growing and harvesting of barley, is 'malting' the barley. The barley is allowed to germinate just enough so that alpha- and beta-amylase enzymes are produced. The barley is then 'kilned', that is, heated, to stop the sprouting and to dry the malted barley, which is now called 'malt'. (Different degrees of heating produce different types of malt ranging from a pale brown to a dark chocolate colour.) When the malt is placed in water heated to between 60° and 70° the alpha- and beta-amylase enzymes break the starch down to produce sugars. This is known as 'mashing' and the mixture is known as the mash. Once all convertible starches have been broken down (this is confirmed by a simple test with iodine), the liquid is then drained off (and

[1] The reader should be aware that there is a great variety of conflicting terminology and notation in this area. For example, $p(h|e)$ is sometimes called the prior conditional probability, with the term posterior probability being reserved for the quantity $p'(h)$, which we will encounter in the next paragraph.

is now known as the 'wort'). The wort is then boiled and cooled. Yeast is introduced which consumes the sugar in the wort, producing alcohol. (Any good introduction to beer brewing will provide more details of this fascinating process—I use Wheeler 1990.)

The amount of alcohol produced during the fermentation process is proportional, *ceteris paribus*, to the amount of sugar in the wort. If we measure the amount of sugar in the wort and in the final fermented beer, we can get a measure of the amount of alcohol by volume in the beer using a simple formula. Therefore, it is important for a careful brewer to measure how much sugar is contained in the wort. This can be done by determining the specific gravity (relative density) of the wort, that is, the density of an amount of the wort divided by the density of an equal amount of plain water. (Density being, of course, mass per unit volume.)

To measure the specific gravity we usually use (in a home-brewery setting) a hydrometer. The following account is based on La Pensée 1990. He first notes that Archimedes discovered that a body floats when it displaces an amount of liquid whose mass is equal to that of the body. (This is why steel ships float: they may be heavy, but the amount of water they displace is heavier.) From this

> [i]t follows . . . that the denser the liquid, the better things float in that liquid. The hydrometer uses this principle to measure the density of liquids. It is essentially a glass tube whose bottom end is loaded with lead shot. The top end of the tube is calibrated with a scale from which the relative density can be directly read by noting how far the hydrometer sinks into the liquid being tested.
>
> (La Pensée 1990: 71)

Hydrometers are marked with gradations so that the depth to which they sink can be measured.

Hydrometers can give a quite accurate measure of specific gravity. There are some caveats, however. Hydrometers are calibrated at a certain temperature: home brewers usually use charts to read the hydrometer at other temperatures. There are a profusion of scales associated with home brewing and the determination of density, which can lead to confusion. Usually these caveats are not important, but Prokop acquired a cheap hydrometer, 'cheap' being the operative word. It seems to have been shoddily put together, and he has doubts as to its accuracy. Still, how hard can it be to make one of these things? So, in the end, there is some, but not a great deal of, uncertainty as to the evidence of the hydrometer reading. (I was going to include a photograph of my cheap hydrometer. Instead I got a perfect illustration of how much can go wrong in experimentation: the plastic container that came with the hydrometer was, bizarrely, nearly opaque, the cheap plastic then fogged up, it was nearly impossible to find a level surface and so the hydrometer stuck to the side of the accursed plastic container, the face of the hydrometer with the numbers inevitably turned away from the camera, and on and on . . .)

Prokop aims to produce a wort with a specific gravity of 1.045 (the details of the scale are not necessary for this example). For concreteness' sake, let us take h to be the hypothesis that the specific gravity of our wort is 1.045, and $\neg h$, of course, that it is not. Evidence e will be that the hydrometer reads 1.045.

Prokop has been very careful in the mashing process. He is familiar with the malt he is using (it is not too old, and has been stored in a cool dry environment), he has carefully regulated hygiene, has been very careful in his measurements, and so on. So, he is quite sure that the specific gravity of the wort is in fact 1.045. He would be willing to wager money on it, and so he can set $p(h) = .95$. It can be easily proved that $p(\neg h) = 1 - p(h)$. Hence, $p(\neg h) = .05$. The question now arises of how likely it is that the hydrometer would give a reading of 1.045 if that is in fact the specific gravity.

Let's take $p(e|h) = .9$, which reflects his relatively low uncertainty in how much his ancient hydrometer will actually confirm the hypothesis, and $p(e|\neg h) = .1$, reflecting the chance that he could get the reading in error.[2] Suppose that e does turn out to be true. Then using Bayes's theorem we can see the degree to which h is confirmed.

Using the theorem of total probability, $p(e) = p(e|h)p(h) + p(e|\neg h)p(\neg h) = (.90)(.95) + (.1)(.05) = .86$. We can now work out that $p(h|e)$ is over .99. (The reader could take it as an exercise to work this out using Bayes's theorem.)

3.3.3 Disconfirmation

Suppose, however, that e turns out not to be true, that the hydrometer gives some other reading. Then we need to calculate $p(h|\neg e)$, once again using Bayes's theorem. We need values for $p(\neg e)$ and $p(\neg e|h)$. Since $p(\neg e) = 1 - p(e)$, $p(\neg e) = .14$. It can also be proved that $p(\neg e|h) = 1 - p(e|h)$, and so $p(\neg e|h) = .1$.

[2] The reader should note that $p(e|h)$ and $p(e|\neg h)$ *need not* add to 1.

So, $p(h|\neg e) = (.1)(.95)/.14 = .67 \ldots$, a radical drop. Still, while h has been disconfirmed, Prokop is more certain than not that the wort does have a specific gravity of 1.045. (You might want to work out what values would lead to $p(h)$ being less than .5.)

The accounts of confirmation and disconfirmation are perhaps unremarkable. This is of course a good thing if we wish to give an account of the relation between hypotheses and evidence. Still, the test of the Bayesian account comes in more difficult scientific episodes.

3.3.4 Am I this good a brewer?—falsification

Prokop quickly becomes an experienced brewer. He begins to formulate his own recipes, which meet with much acclaim (especially the Homesick Pilsner). But he runs into a puzzle: his methods seem too successful. To explain: the rate of extraction of fermentable sugars from malt depends on many factors such as the variety of barley, how well it was malted and kilned, and the conditions under which it has been stored. The latter conditions also depend on when the grain was milled (if freshly milled, then it is less sensitive to storage conditions). The efficiency of extraction is the percentage of the possible maximum extraction of fermentables. For example, under ideal conditions, for a kilogram of pale malt (lightly kilned malt) one litre of wort would have a specific gravity of 1.297 and for flaked maize (a cheap fermentable often used in mass produced American beers) a specific gravity of 1.313. Wheeler informs us that commercial breweries get an efficiency of about 85 per cent, while home brewers will get efficiencies ranging from 80 per cent for strong beers to 90 per cent for weak beers. (All this information is taken from Wheeler 1990: 125–7.) Prokop, however, finds that after measuring specific gravity and calculating he is getting extraction rates above 95 per cent for weak beers. He wonders: 'Am I really this good a brewer?'

Let h be the hypothesis that Prokop's extraction rate will be less than 90 per cent. Prokop has used Wheeler as an authority for a long time, and always found him accurate. On the basis of this information, he assigns a high probability to h, say .90. On the other hand, we have a rival hypothesis: that Prokop is indeed a very good brewer. As we noted before, he is very careful, and very familiar with his malt. Still, he gives Wheeler's pronouncements greater weight.

Prokop has now bought his own hydrometer from a reputable scientific instruments supplier. He thinks that hydrometer is very accurate, and that it could not give the reading that it does if the specific gravity is different. But, his hydrometer reading (which he calculates to be 1050) implies that his extraction rate is greater than 95 per cent. He calls the evidence of the hydrometer e, and assigns it probability .95. Obviously, if e is true, then h is false, so $p(h \mid e) = 0$.

This is a particular case of disconfirmation: falsification. A hypothesis is falsified if it is completely disconfirmed by the evidence. When the probability of $p(h \mid e) = 0$, the occurrence of e means that the probability of h will be 0, no matter what its prior probability. (This can easily be seen by using Bayes's theorem.)

3.3.5 Am I this good a brewer?—The Duhem-Quine problem

But are we really sure that our evidence is so good? We could check to see if the hydrometer is in fact calibrated by dissolving a known amount of sugar in a known amount of water at a given temperature. But then we might also check to see whether our thermometer works, whether the water is pure, whether the sugar is pure, and so on. This gives us a simple domestic example of a typical sceptical move. It seems always possible to deny some evidence by pointing to further (albeit often fairly trivial) unknowns. And these unknowns seem to multiply infinitely.

Indeed, we do often assume a wide range of background knowledge when undertaking investigations: that we are not hallucinating, that we are not being systematically tricked by our colleagues, that our apparatus is operating within the tolerances set for it, and so on. In practice we do not question these assumptions, except in fairly extreme circumstances. Background assumptions are often called *auxiliary hypotheses*.

The question of when to test auxiliary hypotheses has been vexing for some accounts of scientific methodology, in particular for those accounts that see the link between hypotheses and evidence as deductive (hypothetico-deductive, or HD, accounts). It is easy to see why: if some theory h along with auxiliary hypothesis (or hypotheses) a implies evidence e, and if e turns out to be false, then logic tells us that h or a must be false. But logic does not tell us which is false.

The philosophical implications can be significant. For example, Quine often refers to this problem as a reason to embrace holism. His oft-quoted epigram 'our statements about the external world face the tribunal of sense experience not individually but only as a corporate body' (Quine 1953: 41) sums up this view nicely (this problem is known as the Duhem-Quine problem, acknowledging an earlier contribution by Pierre Duhem). A very brief explication of Quine's use of the Duhem-Quine thesis can be found in Appendix A.7.

The Duhem-Quine problem also poses quite a problem for Popper's account of scientific methodology. According to Popper, theories are scientific if they are falsifiable, that is, if there is some evidence that can prove them false. But it seems that no theory is in fact falsifiable due to the Duhem-Quine problem—the evidence can always be discounted by blaming the auxiliary hypotheses. Therefore, it seems, no theory can be scientific by Popper's standard. The solution would be to find a principled way to apportion blame for experimental falsification. Popper and his followers have put forward a number of proposals, none of which have met with much acclaim (some of these are surveyed in Howson and Urbach 1993: 131–6).

Another way of putting the Duhem-Quine problem is that theories do not usually by themselves have empirical consequences. We have already seen that our theory about our brew needs additional assumptions if it is to be testable. For example, Newtonian mechanics *by itself* cannot tell us, for example, about the behaviour of planets in our solar system. It only has consequences when we add information about masses, positions, and momentums. The Duhem-Quine problem is therefore a serious one for any account of scientific theories that cannot distinguish between the differential impact of evidence on a hypothesis and on auxiliary assumptions.

3.3.6 The Bayesian account of the Duhem-Quine problem

We have already seen that Bayesian methodology can explain disconfirmation and confirmation, and so encompasses an HD account of methodology. It can also explain away the Duhem-Quine problem, according to its proponents. We can explain why we normally don't pay attention to background assumptions if those assumptions are considered questionable. Suppose we explicitly represent the auxiliary assumptions as a. When we test some hypothesis h we may take these assumptions to be fixed, that is, we take $p(a)$ to be 1. If, as it is often reasonable to assume, h and a are independent (see the paragraph after next), then $p(h\&a \mid e) = p(h \mid e)$.[3]

[3] To prove this use Bayes's theorem, equation 2 in Box 3.1, and the assumption of independence.

But, sometimes something goes wrong with our equipment, and we would like to see the differential impact of the evidence on the hypothesis and on the auxiliary assumptions using Bayes's theorem. That the impact can be differential is obvious: $p(h|e)$ need not be the same as $p(a|e)$. (If you need, you may write out Bayes's theorem for them to convince yourself.) Therefore, adverse evidence need not affect both hypothesis and auxiliary assumptions the same.

But let's bring this down to earth by visiting Prokop again. First, we reiterate the lexicon.

> e – the specific gravity is indicated to be 1.035
> h – the extract rate is 90 per cent or less
> a – the hydrometer and thermometer are calibrated; tables are correct; not hallucinating; etc. . . .

(a starting specific gravity of 1.035 should result in a light—weaker—beer.) e and a together are incompatible with h.

We make the simplifying assumption that e and a are independent (which is often the case in Duhem-Quine problems). In Prokop's case, that the specific gravity is a certain value does not influence the calibration of my thermometer. If my thermometer is miscalibrated, it is not because I have brewed strong beer.

The crux of the problem is that h and a cannot be jointly true if e is true, that is, $p(h\&a|e) = 0$. But, which one is false? Box 3.1 shows how to derive the values of $p(h|e)$ and $p(a|e)$. We can see that we need values for $p(h), p(a)$, $p(e|h\&\neg a), p(e|\neg h\&a)$, and $p(e|\neg h\&\neg a)$.

Prokop has bought his equipment from a well-known scientific instrument maker, and they are in good shape. He has no reason to doubt that they are well calibrated. Prokop does not take hallucinogens, and has no other strange feelings. Therefore, the prior probability of his auxiliary assumptions concerning the accuracy of his equipment (and state of mind) is very high. Let

$$p(a) = .99.$$

And as before, he is fairly sure that the hypothesis is true, that is,

$$p(h) = .90.$$

Suppose that the equipment works, and that the hydrometer gives a reading of 1.035, and the extract rate is in fact over 90 per cent. Suppose the equipment works, *and* that the alternative hypothesis, that Prokop is a good brewer, is true. There is nothing extraordinary about the reading, then. In fact, we expect it, and so

$$p(e|\neg h\&a) = .9$$

Now suppose that the equipment is miscalibrated, *and* the extract rate is above 90 per cent. The crucial factor is the miscalibration. If the equipment is miscalibrated, we don't expect any particular direction of bias with respect to the hypothesis. It might give readings above, or it might give readings below, say, 1.035. In fact, there is a range into which the reading might fall. Still, this range is limited: ranging from around, say, 1.025 to 1.040. The gradations are in degrees, giving us a range of about 25 possible different values, given the amount of malt used. We will assign these equal values, although we need not, as we shall see in Chapters 5 and 6. Then the probability would be

$$p(e|\neg h\&\neg a) = .04$$

The last case to consider is when our equipment is in fact miscalibrated, but the extract rate is the expected 90 per cent. Employing exactly the same reasoning earlier we can set

$$p(e|h\&\neg a) = .04.$$

Calculations then show that

$$p(h|e) = .02 \ldots$$

and

$$p(a|e) = 0.977 \ldots$$

The example we have given is a somewhat easy one, since there are only two hypotheses available. This has the effect that $p(e|\neg h\&a)$ has a large value, that is, the alternative hypothesis predicts the evidence—what is lost by h is gained by $\neg h$. If there were no alternative hypothesis (say, Prokop was not very familiar with his malt, and so $p(e|\neg h\&a)$ would not necessarily be large, or perhaps even very small), the disconfirmation of the hypothesis will not in general be so dramatic, since many other hypotheses might share the probability conferred by e. This is, of course, exactly what we should expect.

Box 3.1 Equations necessary for solving the Duhem-Quine problem

(1) $$p(a|e) = \frac{p(e|a)p(a)}{p(e)}.$$

We are given $p(a)$, and so need to determine the values of $p(e|a)$ and of $p(e)$. A version of the theorem of total probability gives

(2) $$p(e|a) = p(e|a\&h)p(h|a) + p(e|a\&\neg h)p(\neg h|a).$$

(We need this version since h and a have no empirical implications alone.) Since h and a are independent and jointly incompatible with e, (2) simplifies to

(3) $$p(e|a) = p(e|a\&\neg h)p(\neg h).$$

We now need to solve

(4) $$p(e) = p(e|a)p(a) + p(e|\neg a)p(\neg a).$$

We thus need

(5) $$p(e|\neg a) = p(e|\neg a\&h)p(h|\neg a) + p(e|\neg a\&\neg h)p(\neg h|\neg a),$$

which, by the independence of h and a, simplifies to

(6) $$p(e|\neg a) = p(e|\neg a\&h)p(h) + p(e|\neg a\&\neg h)p(\neg h).$$

Substituting (3) and (6) into (4) we get

(7) $$p(e) = p(e|a\&\neg h)p(\neg h)p(a) + [p(e|\neg a\&h)p(h) + p(e|\neg a\&\neg h)p(\neg h)]p(\neg a).$$

Finally, substituting (3) and (7) into (1) we get

(8) $$p(a|e) = \frac{p(e|a\&\neg h)p(\neg h)p(a)}{p(e|a\&\neg h)p(\neg h)p(a) + [p(e|\neg a\&h)p(h) + p(e|\neg a\&\neg h)p(\neg h)]p(\neg a)}.$$

Repeating the exercise for

(9) $$p(h|e) = \frac{p(e|h)p(h)}{p(e)}$$

gives us

(10) $$p(h|e) = \frac{p(e|h\&\neg a)p(\neg a)p(h)}{p(e|h\&\neg a)p(\neg a)p(h) + [p(e|\neg h\&a)p(a) + p(e|\neg h\&\neg a)p(\neg a)]p(\neg h)}.$$

However, there is room for doubt that this is a solution, depending on what you count as a solution (and, indeed, what you count as a problem). Different values will give different results (and this can be taken as a problem with Bayesianism, one of the main themes in 3.8 and 3.9). A spreadsheet can be found at <http://probability.cz> which will assist with the relevant calculations. The reader is invited to enter their own values into the spreadsheets, or to invent, and solve, Duhem-Quine problems of their own.

Key readings on the Bayesian approach to the Duhem-Quine problem are Dorling 1979, Redhead 1980, Howson and Urbach 1993, and Jeffrey 1993. Recent (advanced) discussions include Bovens and Hartmann 2003: 107–11, Strevens 2001, and the ongoing exchange between Fitelson and Waterman 2005 and Strevens 2005a.

3.3.7 Other Bayesian solutions

We can now sum up the Bayesian account of relations in Box 3.2. These relations have been used to propose Bayesian solutions to a number of long-standing problems in the philosophy of science. I refer the interested reader to the bible of such matters, Howson and Urbach 1993, especially chapter 7.

Another area of interest that has recently exploded is quantitative Bayesian confirmation theory. We have only discussed the comparative notions of confirmation and disconfirmation. Bayesianism, however, is a quantitative theory, and so it seems that we might be able to also obtain a theory of degree of confirmation. The recent debate in this area was kicked off by Milne 1996. Christensen 1999 and Eells and Fitelson 2000 are also key contributions. Fitelson's 2001 doctoral thesis contains a useful history and discussion of the field up to 2000. The debate sheds light—or casts doubt—upon the Bayesian solutions touted by Howson and Urbach.

After many years in the wilderness, Bayesianism's popularity has exploded, and at the time of writing it is the dominant account of learning from experience in the philosophical literature (but not yet in the statistics literature). Still, there remain doubts, to which we now turn.

Box 3.2 Bayesian confirmation relations

Confirmation Relations

Verification

Prokop flips the light switch; the light comes on. He takes the proposition that the light bulb is not burnt out to be true.

Falsification

Prokop flips the light switch; the light does not come on. He takes the proposition that the light bulb is not burnt out to be false.

Confirmation

Prokop flips the light switch; the light comes on. He takes the proposition that the light bulb is not burnt out to be more likely to be true.

Disconfirmation

Prokop flips the light switch; the light does not come on. He takes the proposition that the light bulb is not burnt out to be less likely to be true.

Disconfirmation of auxiliary hypothesis

Prokop flips the light switch; the light does not come on. He checks to see if lights are working elsewhere, discovers they are not, and concludes that the electricity in the house is not on. He takes the hypothesis that the light bulb is not burnt out to retain the same likeliness; he takes the auxiliary hypothesis that electricity is working in the house to be disconfirmed.

Disconfirmation of main hypothesis

Prokop flips the light switch; the light does not come on. He checks to see if lights are working elsewhere, discovers they are, and concludes that the electricity in the house is on. He examines the light switch and sees that it is in working order. He takes the hypothesis that the light bulb is not burnt out to be disconfirmed.

Bayesian Account

Vocabulary

The light switch is working $= h$
Upon flipping the light switch, the light comes on $= e$
Electricity in the house is on, and the filament is not broken, etc $\ldots = a$
p the initial probability distribution, p' is the probability distribution after learning e.
(For simplicity's sake, probabilities are assumed to be non-zero.)

Verification

$p(h|e) = 1$, so since $e, p'(h) = 1$.

Falsification

$p(h|\neg e) = 0$, so since $\neg e, p'(h) = 0$

Confirmation

$p(h|e) > p(h)$, so since $e, p'(h) > p(h)$

Disconfirmation

$p(h|\neg e) < p(h)$, so since $\neg e, p'(h) < p(h)$

Disconfirmation of auxiliary hypothesis

$p(h\&a|\neg e) < p(h\&a)$, where $p(a|\neg e) < p(a)$ and $p'(h) = p(h)$, so since $\neg e$, $p'(a) < p(a)$.

Disconfirmation of main hypothesis

$p(h\&a|\neg e) < p(h\&a)$, where $p(h|\neg e) < p(h)$ and $p'(a) = p(a)$, so since $\neg e$, $p'(h) < p(h)$.

3.4 Problems with the Dutch Book Argument

There are two main categories of worries about Bayesianism. The first is with the foundations of Bayesianism. These concern, for example, problems with the interpretation of the Dutch Book argument, its validity, and the scope of its application. The second class of problems concern questions as to the adequacy of the Bayesian account as an account of science (and of learning from experience) in general. The concern is whether or not Bayesianism adequately describes what scientists, or at least reasonable people, do. In this section we will discuss the former kind of problem; in section 3.8 we will discuss one example of the latter.

The main foundational problem with the Dutch Book argument is the question of how it should be understood. One reading that can perhaps be ascribed to de Finetti (the first along with Ramsey to make the argument explicit) seems to be that it should be read literally. If someone does not follow the probability calculus in explicit assessments of uncertainty in terms of bet, then they are in danger of being made to lose that bet. At the other end of the scale are those (perhaps also including de Finetti) who see the Dutch Book argument as a dramatic tale that does not entail that any actual set of beliefs will lead to any actual consequences. After detailing these two extremes, I will address a third interpretation which claims to leave this scale aside. (This survey of the Dutch Book argument follows the structure of Childers 2009.)

3.4.1 The literal interpretation of the Dutch Book argument

Earlier we discussed a willingness to bet at certain odds as indicative of the degree of belief. But there are many people, myself most emphatically included, who do not like to bet. Such an aversion to betting can skew the odds that we offer. Thus, bets may not, and for me most certainly are not, related to actual degrees of belief. (This objection is probably put forward most forcefully by Schick 1986. See also Armendt 1993.) Consider the Dutch Book argument for the third axiom: this in fact requires the assumption that the value of money is additive. Suppose that I have set aside a portion of my limited income and am considering buying a bet on A, which I believe quite strongly, but not with certainty, to be true. If I were to only buy the bet on A, then I would pay a high price, that is, I would give a high betting quotient. But now suppose that I have the opportunity to buy a bet on a proposition B, which is mutually exclusive with A. It seems natural that I would, in this circumstance, be less willing to pay the same amount for A if bought as part of a package of bets than I would be willing to spend on A if bought alone (to minimize the strain to my budget). This seems to be how most people feel: they wish to pay less for items bought in bulk. And the more we gamble, for those of us who are risk averse, the less risk we wish to expose ourselves to.

Schick labels the assumption that people are willing to offer the same betting quotients singly that they would offer jointly 'value additivity'. Another, more traditional, way of putting the objection is that the value of money does not add up in a simple fashion. And yet, as has been pointed out on many occasions, the Dutch Book argument assumes that the value of money does add up simply. This criticism seems to me correct: the fair betting quotient p is a ratio derived from a particular betting situation in which the bettor is willing to risk forfeiting some amount of money b to get an amount a if the proposition being betted on is true. It makes no sense to assume that this proportion will be the same no matter the size of a and b. (This criticism also has implications for the Dutch Book argument for the fourth axiom.) Most people would offer much more conservative rates the larger the amount of money involved. It seems, then, that the Dutch Book argument rests on a demonstrably false assumption about how the value of money adds up.

One standard response to this objection is that the argument only requires that we use small amounts of money to bet relative to our total fortune, and so the distorting effects of the non-linearity of money are minimized. This, of course, requires that we develop an independent theory of the relation between value and money—no trivial task. But even granted that it could be done, another problem arises: if the amount of money is too small the bettor may not be motivated to carefully give odds (after all if the stake of a bet is 50 Czech hellars, a tiny currency unit now demonetized, it doesn't matter too much if you win or lose). But then if the bet is too big, you will be motivated to be conservative. In neither case do your bets show what your degrees of belief are. Thus, we would need to find the magic point where bets do match degrees of belief. But if we are betting over many different propositions, we may find that there is no such magic point: the bets on the individual propositions may be too small, the bets on the joint propositions too large. There is thus no straightforward link between overt behaviour and beliefs.

Another possibility is to use the device of a lottery to construct a currency denominated in units of utility. Savage in, for example, 1971, following Smith 1961, refers to the construction of such a currency. We find some mechanism which our subject is willing to say is fair. Let us use the example of a lottery: a fair lottery is one in which the participant thinks that the likeliness of any particular ticket being drawn is the same. We can then take the prize of the lottery to be equal to the value of all the tickets, and the tickets can then serve as units of currency in a bet. (Think of the extreme case: you have all the tickets in the lottery, and so you will win. Therefore the tickets together have the same value as the prize. Note, however, that the assumption that the value of the tickets divides evenly once again rests on some fairly strong assumptions.) The basic assumption would be that we can (always) find such a lottery or other chance mechanism. I think we can, and it is a point I shall return to. But a utility currency still won't mollify critics of the betting approach to eliciting probabilities.

One such critic is Prokop's neighbour, the Reverend. He's a Protestant, but not like the ones at home. His intense gaze makes Prokop very nervous, and the Reverend seems oddly drawn to the topic of why Prokop has a pair of pink shoes. The Reverend doesn't drink, and he doesn't gamble. He thinks *Dungeons and Dragons* is a satanic game, and that homosexuality should be a crime, something he often mentions while switching his glare between Prokop's face and shoes. Prokop one day innocently mentions that he's been studying the relationship between probability and betting, which sparks off a furious five-minute sermon on the evils of gambling.

It's pretty clear that no one is going to be eliciting betting quotients from the Reverend. Therefore, there is no link between betting and the Reverend's degrees of belief, because he does not bet. One traditional way around this broken link would be to forge it: to compel the Reverend to provide betting quotients. The thought of this makes Prokop shudder, because he has a good idea of just how willing the Reverend would be to be a martyr. Perhaps we might want to exclude counterexamples to the link between belief and behaviour such as the Reverend. But then we would have to offer a substantive theory of what a reasonable bettor looks like: and this is just what our theory was supposed to solve. Therefore we should come up with *independent* grounds to dub the Reverend unreasonable. This detracts considerably from the simple empiricist picture of our Dutch Book argument.

I leave aside the practical details: are we to rob the Reverend so that he may provide his share of the stake? Perhaps we could take *his* shoes. If we take a bet as a contract, which seems reasonable, is it still valid, since it is coerced? Hobbes might say yes, since he argues in *Leviathan* (chapter 14) that 'Covenants extorted by fear are valid'. And how could we possibly get odds from the Reverend? If he chooses to be tight-lipped, or to only recite Bible verses, there is no way we can get assent or dissent from taking any particular odds on the bet. (Milne 1991 argues that Bayesianism actually *requires* those forced to bet to give odds not in accordance with their degrees of belief.)

Finally, one possible response would be that we are dealing with an idealization, and that such failures of actual behaviour need not be of great importance for the argument. But the idealization is not a natural one: why should the hero of our story about Dutch Book arguments have inclinations to bet in just such a way as to make the argument work? This response is as question-begging as the previous: why this idealization, and not another?

3.4.2 The as-if interpretation

The forgoing interpretation of the Dutch Book argument requires that people actually be willing to place money for a bet, instead of merely naming some quantity which they think is a fair price. This is overly behaviourist: thoughts and behaviour are not always so closely linked, even if there is *some* link between behaviour and belief under some circumstances. Howson and Urbach 1993 suggested that a better interpretation of the Dutch Book argument would be a counterfactual one: the betting rate is the rate that one *would* be prepared to offer if one *were* to bet:

> Attempts to measure the values of options in terms of utilities are traditionally the way people have sought to forge a link between belief and action, and much contemporary Bayesian literature takes this as its starting point. We do not want to deny that beliefs have behavioural consequences in appropriate conditions, they clearly do, but stating what those conditions are with any precision is a task fraught with difficulty, if not impossible . . . [T]he conclusion we want to derive, that beliefs infringing a certain condition are inconsistent, can be drawn merely by looking at the consequences of what *would* happen if anyone *were* to bet in the manner and in the conditions specified.
>
> (Howson and Urbach 1993: 77)

This interpretation seems superior to the literal one. It appears to avoid problems about the additivity of money, or of unwilling subjects who object to betting since no money actually changes hands. The Dutch Book argument is, under this interpretation, conceived of as a thought experiment, establishing rules of consistency of ideal behaviour.

The question then arises how to interpret the as-if interpretation. The natural way would be to read willingness to bet in a counterfactual, or subjunctive, manner. And there is a standard semantics for dealing with these counterfactuals, the Lewis-Stalnaker semantics. Unfortunately, this interpretation of the Dutch Book argument makes it invalid. This can be seen by formulating the Dutch Book argument for the third axiom as follows.

(1) If you were to bet on A you would regard p as a fair betting quotient.
(2) If you were to bet on B you would regard q as a fair betting quotient.

Therefore

(3) If you were to bet on A and on B, you would regard p and q as fair betting quotients.

This is an instance of the so-called counterfactual fallacy of strengthening the antecedent, that is, the argument from 'If A were the case then C' to 'If A and B were the case then C'. For example, suppose the counterfactual conditional 'If I were to run to the bus stop I could catch the bus' is true. But if the bus is early, or if it is cancelled, or if it has been sucked into a black hole, then even if I do run, I won't catch the bus. (This argument was suggested to me by a similar argument in Anand 1993. A discussion of strengthening the antecedent of counterfactual conditionals can be found in Lewis 1973: 17.)

It might be suggested that the argument could be formulated in another logic: counterfactual conditionals with Lewis-Stalnaker semantics just might not be the correct way to formulate the argument. But it seems that any logic which plausibly formulates the 'as if' Dutch Book argument will also show the argument to be invalid. Consider Harold, one of Prokop's American friends, known to all as Dirty Harry (and deservedly so, given his hygienic proclivities). Dirty Harry lives in Prague, and it is summer. He is feeling suicidal (he lives in Prague, which is great, but no girls will approach him), and so is disposed to kill himself. But, it's hot, and he is also disposed to have a beer. Suppose he drinks the beer first: then perhaps he no longer feels suicidal, because the beer is so good, or perhaps because he gets too drunk to remember his troubles. Or perhaps, being so drunk, he accidentally stumbles in front of a tram, and is run over before he can kill himself by his own hand. Conversely, suicide precludes beer drinking. So, Dirty Harry's beer-drinking disposition blocks his disposition to commit suicide.

The counterfactual approach was introduced to dispense with these interfering factors like the non-additivity of the value of money. But any logic which faithfully represents dispositions to behaviour will also represent dispositions in general not being serially or jointly realizable. And this means that the Dutch Book argument is invalid, since it assumes that dispositions *are* so realizable. Thus, it seems, this interpretation will not do either. (I should also point out that Colin Howson has withdrawn his support for this interpretation in favour of the next to be discussed.)

Is the Dutch Book argument doomed? In either of the two interpretations discussed so far, I think it is. Still, there is a possible spectrum of interpretations of the Dutch Book argument, with appeals to actual behaviour at one extreme and considerations of what-if at the other. Both of the extremes are not sufficient to ground the argument. So perhaps there is a middle ground, although it's hard to see what it would be like.

As well, both interpretations require a link between belief and action—they require that belief be a disposition to action. This is typical of a pragmatist (and empiricist) view of belief. In 1927 Ramsey took the view that the meaning of a sentence was in some way a disposition to action. Quine in the entry 'Belief' in his dictionary *Quiddities* (1990) also takes belief to be a disposition to action, and explicitly links this view with the behaviourist interpretation of the Dutch Book argument. But we have seen that the link between disposition and action in the case of betting is so weak that it seems that such an identification is pointless: we might *say* a belief is a disposition to action, but we may never be in a position to establish that it is.

The two interpretations we have discussed so far try to save what might be thought of as an empirical interpretation of the Dutch Book argument. One last attempt to save this interpretation would be to jettison the notion of belief completely, and just talk in terms of penalties (in terms of, say, lost bets, or scoring rules; see Appendix A.3.2) and the probability calculus. But a straightforward linking of adherence to the probability calculus and consequences of that adherence is always doomed to failure, for the same reasons that the attempt to link logic and behaviour are doomed: if someone behaves illogically, say, believing contradictions, they will not necessarily be struck down by lightning bolts hurled by an angry god of logic. There will always be circumstances where inconsistency is harmless (or perhaps even helpful). Therefore, at least after Frege, logic is justified not by a crude appeal to actual or imagined consequences, or to psychological states like judging, but rather in terms of its adequately capturing some relevant feature of (again, idealized) reasoning. This leads up to our next interpretation of the Dutch Book argument.

3.4.3 The 'logical' interpretation

One interpretation of the Dutch Book argument that has recently received much attention is that the probability calculus serves to provide a notion of *consistency* of belief in a purely *logical* sense. (This can be contrasted to conditional probabilities providing a generalized notion of *consequence* in Chapter 5, hence the 'logical' in the title of this subsection.) This idea can be found in de Finetti and Ramsey: a new way of explicating this idea has recently been put forward by Colin Howson in a number of papers (for starters, I can recommend Howson 2003 and Howson and Urbach 2006).

In what follows we will stick with a (vanilla formulation of) a classical logic (not much hinges on this). A truth valuation for a set of sentences is consistent if it can be extended to cover all sentences in the language in accordance with the basic semantic definition, which lays down rules for truth assignments. To use Howson's example, if we were to evaluate $A \rightarrow B$ and A as true, but B as false, there would be no assignment of truth values to all the other sentences of the language in accordance with the basic semantic definition. As he puts it, the problem is to solve a system of equations where we are given $v(A \rightarrow B) = 1$, $v(A) = 1$, but $v(B) = 0$. There is no such solution to such a system of equations of truth assignments, and so it is inconsistent. This also tells us what is deducible from $A \rightarrow B$ and A, namely, all those sentences X for which $v(\text{not-}X) = 1$ cannot be consistently assigned if $v(A \rightarrow B)$ and $v(A) = 1$, and so we get the notion of consequence as well.

According to Howson, the parallel notion of consistency for probability is that of assignments of fair betting quotients to propositions. An assignment of fair betting quotients to a set of propositions is consistent if it can be extended to an assignment over all propositions. The betting quotients can only be fair if (and only if) they obey the rules of the probability calculus. The fair betting quotients are the semantic objects of our language, the analogue of truth values. The syntax is simply the probability calculus. Thus, the Ramsey-de Finetti theorems serve as a kind of soundness and completeness theorems: they show that the syntax—the probability calculus—and the semantics—fair betting quotients—are in complete agreement.

Hence Howson holds that he is offering a logical reading of the Bayesian interpretation: how to consistently assign numbers to propositions. These numbers may serve as a heuristic or as an explication of certain aspects of uncertainty. For example, we could take them to be betting quotients. But, like the notion of truth in logic, the notion of uncertainty in Bayesianism remains, according to Howson, to a large degree independent of a substantive theory of uncertainty.

This interpretation of the Dutch Book argument of course severs any link between behaviour and realized belief on the one hand and the probability calculus on the other. But this seems a high price since Bayesian epistemology should be about belief, and beliefs are valuable because they either fit the world or not, and so guide our actions. In particular, it seems odd to talk about fair betting quotients, and then deny a link to behaviour. There is a very tight link between truth and consistency in the sense that consistent valuations capture something important about the notion of truth. The link between degrees of belief and betting quotients is not as apparent. This is perhaps because at bottom belief is an epistemic notion, and we are interested in it for pragmatic reasons. Truth remains a metaphysical notion that is easily separated from consequences. If we remove the link between actual betting and actual belief, and fair betting quotients, we may ask what Bayesianism has to do with epistemology. (Of course, someone might argue that truth is not a metaphysical notion. But a reduction of truth to a pragmatic or epistemic notion would simply lead to the problems of the behavioural reading of the Dutch Book argument.)

There is another way of linking degrees of belief and probabilities that does not involve betting. This reading also seems to fit better with Howson's interpretation of the Dutch Book argument, and so we will cover it in the next section.

3.5 Probability from Likelihood

Prokop has won some bets, and lost some bets. He doesn't like losing bets, and, he discovers, he's not really attracted to gambling at all. He finds it difficult to think in odds, and then there are always practical concerns with betting anyway: how much is too much, or too little? How much does the person arranging the book get? And there are theoretical concerns: if he can't think in bets, is he really giving his true assessment of likeliness?

There is a television show that Prokop quite likes: 'Wheel of Fortune' is centred around a wheel of fortune, oddly enough, spun by players, with moves of the game determined by which section is picked out by a fixed pointer when the wheel comes to a stop. He also has taken to sitting in random lectures (naturally, with the permission of the lecturer). One was a psychology course on the elicitation of uncertainty. This confluence of happenstance led Prokop to construct his own wheel. The simplest version is a clever construction: it is a wheel on which he can set the colour of an

Figure 3.1 A probability wheel

arbitrary sector. He has taken to tormenting his friends. In Figure 3.1, the portion in dark grey represents how likely it is that it will rain tomorrow. Will it be more, or less, likely than as depicted?

With repeated questioning, he usually finds his friends, if they sit still long enough, settle down to a particular segment. He discovers, browsing through old articles, that this can be used as the basis for deriving a 'representation theorem'. (Arguments to the effect that uncertainty, taken as degrees of belief, can be *represented* as probabilities are called, not surprisingly, representation theorems.)

We will now proceed step by step through the representation theorem. Since we've already had a dose of algebra in section 3.2.4, we'll relegate this section's algebra to Appendix A.6.3. The first step in the representation theorem is to order our propositions in terms of likeliness. We do not mean 'probable', but something much less sophisticated: is it more likely, say, that it will rain tomorrow than not? Is it more likely that you will be struck by a meteorite than get a decent beer in Lake Charles, Louisiana? This means that likelihood is a primitive notion in this representation. Although it can be defined in a number of ways, we shall not do so, and assume that the primitive notion is clear enough. We will now explicate its properties. (Likelihood will be defined over a field of propositions, as with the Dutch Book argument.)

We will assume that likelihoods obey the axioms of *qualitative probability*. These are three. First, we assume that we can order propositions by how likely we think they are (for any two propositions, one is likelier than the other, or they are equally likely). Second, we assume that if for any two propositions, if one is at least as likely as the other, it will remain so even when another proposition contradicting both propositions is taken under consideration. To use an example from French (1988: 227), if I think that it is at least as likely that the next roll of a die will be a five or a six as a one, then I should think it at least as likely that the next roll of a die will be a five, six, or two as a one or a two. Thirdly, we make a technical assumption that the certain event is more likely than the impossible event, and that every event is as least as likely as the impossible event.

These axioms are (relatively) uncontroversial. It would be nice—indeed it would be *very* nice—if we could show that every likelihood ordering determines a corresponding probability relation that preserves the likelihood relation, and vice versa. That is, we would like to show that *A* is more likely than *B* if and only if *A* is more probable than *B* for any two given propositions *A* and *B*. This, however, cannot be shown, since there is no such correspondence, as Kraft, Pratt, and Seidenberg 1959 proved. (They showed that there is a qualitative probability ordering of five propositions that have a corresponding probability distribution that does not preserve this ordering. Fishburn 1986 provides a survey of research on qualitative probabilities.) We will therefore require some additional constraints. In this section we will use a representation that allows us to visualize uncertainty in such a way that we get the desired correspondence between the likelihood ordering and the probabilities. (In section 3.6 we will use lotteries to get this correspondence.)

We will call our (visualizable) representation of uncertainty, following French, a *reference experiment*. Although nothing turns on which representation we use, we shall use the wheel of fortune (or probability wheel): a disk with a (perfectly balanced and oiled) spinning arrow. (You should keep in mind that the reference experiment is needed to provide additional structure so that likelihoods and probabilities match up.) The arrow's landing at some particular point is an event. We then construct a field of events, which will contain points, intervals, and combinations of intervals.

There is an obvious notion of likeliness related to the length of intervals: for any two events, one is likelier than the other if the sum of the lengths of its interval (the arc on the circumference of the circle) is longer (see Figure 3.2).

Here *A* is more likely than *B* as well as *C*, while *B* and *C* are equally likely. *A*, *B*, and *C* are all also mutually exclusive: if one happens, the others will not. Length obeys the axioms of qualitative probability, and so provides us with necessary mathematical structure.

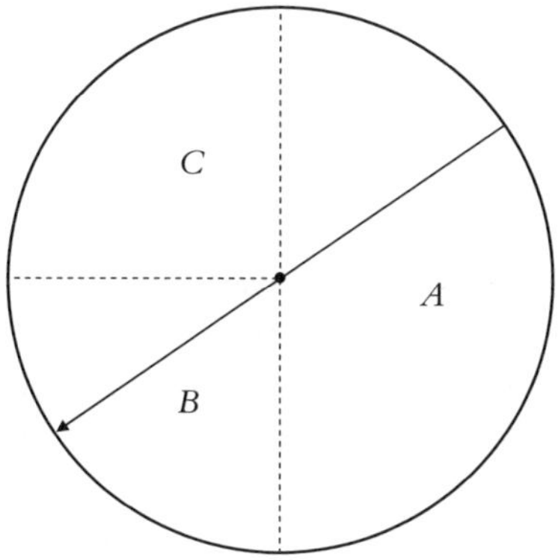

Figure 3.2 A reference experiment

The final step in our construction is to link the events from the wheel of fortune with the events in our original algebra. We match up each event with a point or an interval on the wheel. We equate the certain event with the entire circumference of the wheel, and the empty event with length 0. The percentage of the circumference of the wheel of fortune an event takes up is the probability associated with that event. From Figure 3.2, it's quite easy to see how the probability axioms act.

This shows that, if we are willing to visualize uncertainty with the help of a reference experiment, subject to some very plausible constraints, uncertainty, or degrees of belief, can be represented by probabilities. This representation also provides a means for discussing the interpretations of the Dutch Book arguments. It clearly fits with the 'logical' interpretation. But we could also introduce penalties to attempt to guarantee elicitation of uncertainty by associating prices with the truth or falsity of a proposition proportional to the area on the reference experiment it takes up. Still, this interpretation remains independent of a betting situation.

3.5.1 Problems with probabilities from likelihood

This way of representing uncertainty has not been as popular as the Dutch Book argument: in fact, it is rarely discussed in the literature. There seem to be several reasons why it is not. First, it requires that people have a pre-existing notion of (variable) likelihood available which can be meaningfully applied. This can be doubted. If we follow Ramsey 1926 and others in taking belief as being linked with disposition to action, then the degree of belief may not be open to direct introspection, as is required in this construction. Instead, being put in a betting situation activates the dispositions properly so that the degree of belief can be measured. Secondly, there is no element of compulsion, and so, perhaps, one could lie about one's degrees of beliefs. This is supposed to be discouraged in a Dutch Book situation (ignoring, of course, the question about the non-linearity of the utility of money). Finally, the reference experiment requires a flat distribution: that we can imagine a reference experiment where each point has equal probability. This is seen by some as a very strong assumption.

I do not think that these objections are very strong; however, I should remind the reader that I am very much in the minority. The first and second objections are actually two sides of the same coin: they rest on the requirement that beliefs be actualized in a consistently measurable way. As we already pointed out, this can be done with the representation (although it would then share all the problems of the Dutch Book argument). The last objection seems blunted by the popularity of the interpretations of probability in Chapters 5 and 6 which postulate flat distributions.

3.6 Probabilities from Preferences

Vladimír offers to cook Prokop dinner, either steak or chicken. Prokop doesn't much care for chicken, but he loves a steak, dripping with just warmed blood. Vladimír decides to try an experiment: he offers Prokop the choice of two games: in both, he will flip a coin, which he shows Prokop. In the first game, if it comes up heads, Prokop gets steak, and if tails, he gets chicken. In the second, if it comes up heads, Prokop gets chicken, and if tails, he gets steak. Prokop says he doesn't care which game, and asks if Vladimír could get to work in the kitchen. Vladimír infers that Prokop thinks the coin is fair: that there is a 50 per cent chance of it coming up heads. For if he thought the coin biased, say, towards heads, then Prokop would want the first game.

We can use this simple device to build up a scale of utility. Suppose Vladimír now offers Prokop a range of other possible meals, one of which is his famous Pardubická Paella. If it turns out that Prokop is indifferent between getting Pardubická Paella and playing a game with a 50 per cent chance of steak or chicken, then we can say that Prokop ranks the paella exactly halfway between steak and chicken. Given enough alternatives (and utility theory always assumes that there are enough), we can then find something between Pardubická Paella and steak, and something between

Pardubická Paella and chicken by offering a similar gamble. And then we can find something between that and chicken, and steak, and so on, so that we can develop our scale of desirability.

The numbering of the scales is determined by the top number and the bottom number. For example we could assign steak 10 and chicken 0. Then Pardubická Paella would be ranked 5. The other numbers are then determined with respect to these numbers. The particular numbers used are not important, except for convenience, since the scales can be uniquely translated ('transformed') to each other. This is exactly analogous to the case of temperature, for example, where there are many different scales (Kelvin, Farenheit, Celsius), which can all be transformed to one another.

Now that we have a desirability scale, we can determine arbitrary probabilities of gambles. Suppose that Prokop is faced with the choice of getting a ham sandwich or Southern fried chicken for lunch, to be determined by the role of a 12-sided die (he has many unusual dice lying around his apartment, for playing *Dungeons and Dragons*, of course). Fried chicken is, obviously, superior to a ham sandwich (Prokop shudders). The gamble is: fried chicken if the die is greater than 6, a ham sandwich otherwise. Let us denote this gamble L (for lunch). He assigns some value to L, which we will also denote L (following the long-established probabilist tradition of being sloppy with notation). He has also assigned the value to the gamble C to get fried chicken if the die is greater than 6, nothing otherwise. And he has assigned the gamble H to get a ham sandwich if the die is 6 or less, nothing otherwise.

Denoting p by the probability of the die coming up greater than 6, we can sum up the separate gambles to produce the combined gamble

$$L = pC + (1 - p)H$$

Simple algebra will show that

$$p = \frac{L - H}{C - H}$$

(The particular numbers are not important: the scale we will use is like a temperature scale. All that matters is that different scales can be translated into each other, and that the bigger the number, the more desirable the outcome. As an example, assume that $L = 6$, $C = 9$, and $H = 4$. Prokop doesn't think his die is fair! Remember to bring your own dice when you play *Dungeons and Dragons* with him.) The theory therefore gives us a simultaneous account of both desirability and probability. I must admit that my account is taken entirely from Jeffrey 1983, chapter 3. This is because Ramsey's ideas are not expressed in the most easily accessible manner, and Jeffrey provided a very nice way of working Ramsey's ideas out.

Another way of working Ramsey's (and de Finetti's) ideas out can be found in Savage 1954. Savage lays down certain postulates governing preference that seem to him (and others) reasonable. He obtains probabilities in more or less the same fashion we did in the preceding section, by postulating a uniform probability distribution.

Finally, another account of the relation between utility and probability can be found in von Neumann and Morgenstern's monumental 1944 *Theory of Games and Economic Behaviour.* (A much more accessible account is the classic Luce and Raiffa 1957.) Von Neumann and Morgenstern explain subjective probabilities and utilities with the aid of externally generated probabilities. That is, they show that, given certain assumptions, we can measure utility if we assume objective probabilities. The resulting formal theory shares the important features of Savage's theory.

3.6.1 Problems with utility theory

There are, of course, many problems with utility theory's account of utility and subjective probability. We cannot, and so will not, cover them all. But first, we can observe that utility-theoretic justifications of subjective probabilities do not in fact give a separate account of subjective probability. In the account in the preceding section, p is recovered from preferences; but it is not defined in terms of preference. The probabilities are not assigned, say, by giving faces equal probability, since Prokop doesn't think the die is fair. So we must use some theory to define the probabilities in the gambles.

In fact, both Ramsey and Jeffrey give Dutch Book arguments for using subjective probabilities in our gambles. As already noted, Savage gave more or less the same version of subjective probability. Von Neumann and Morgenstern pursued a similar path, using objective probabilities to determine subjective probabilities. Thus, utility theory inherits all the problems of the previous justifications of subjective probability.

As well, all versions of utility employ a notion of independence in certain choice situations. This is usually in the form of an axiom (or in the form of a theorem that can be derived from the axioms). It says that some alternative A is preferred to another alternative B if and only if you prefer any lottery giving either A or C to any lottery giving B or C. That is, mixing in a third possibility shouldn't rearrange your preferences. Both of the lotteries will with some chance give you *C*. They only differ in whether there is a chance that you will get *A* or *B*. So, if you really do prefer *A* to *B*, you should prefer the lottery that gives you a chance of getting *A*, no matter how small.

This axiom (or theorem) is necessary for a mathematically 'nice' utility theory. It also leads directly to a powerful counterexample to utility theory, the Allais paradox. Suppose that you are offered the following choice. Either you can have

A: $1 million

or the following gamble:

B: $1 million with an 89 per cent chance, $5 million with a 10 per cent chance, or $0 with a 1 per cent chance.

You should make a choice now.

Now suppose that you are offered the choice of the following two gambles:

C: Nothing, with an 89 per cent chance, $1 million with an 11 per cent chance

or

D: Nothing, with 90 per cent chance, $5 million with a 10 per cent chance.

You should again make your choice.

It perhaps comes as no surprise that most people prefer *A* to *B* and *D* to *C*. (If the size of the pay-offs don't give you these preferences, multiply them by some number—10, or 100, or .5, whatever—until you do.) It's worth noting that a minority prefer *B* to *A*, and *C* to *D*. What follows applies to these preferences too. And this leads to a rather large problem: if we are to act so as to maximize our expected utility, these choices are inconsistent. Suppose that we have a utility function *u* that will yield the utility of some amount of money. To get expected utility we multiply this function by the chance of getting the amount of money. From the choices introduced a moment ago, we know that the certain utility of gamble *A* is greater than the 89 per cent utility of $1 million plus the 10 per cent utility of $5 million plus the .01 per cent utility of nothing. That is,

$$u(\$1 \text{ million}) > .89u(\$1 \text{ million}) + .10u(\$5 \text{ million}) + .01u(\$0)$$

Similarly from the choice of *D* over *C* we know that

$$.90u(\$0) + .10u(\$5 \text{ million}) > .89u(\$0) + .11u(\$1 \text{ million})$$

There is, however, no function u which satisfies these two equations, as can be seen by a bit of algebra. Subtracting $.89u(\$1 \text{ million})$ from both sides of the first equation gives us

$$.11u(\$1 \text{ million}) > .10u(\$5 \text{ million}) + .01u(\$0),$$

while subtracting $.89u(\$0)$ from both sides of the second equation gives us

$$.01u(\$0) + .10u(\$5 \text{ million}) > .11u(\$1 \text{ million}),$$

which is a contradiction.

Why did this contradiction happen? Because of the way we can solve the system of inequalities. That we can do so is ensured by the independence postulate. As we noted, the presence of this axiom makes utility theory much more mathematically smooth. It also produces the wrong answers.

One very common response to the Allais paradox is to invoke a distinction between normative and descriptive accounts of decision-making (for example, Savage 1954: 101–3). The claim is that the Allais paradox shows us that expected utility theory does not describe how people actually make their decisions, but that this is irrelevant, since we should be concerned with how people *should* make their decisions (i.e. the theory describes how an ideal decision-maker would decide). But of course, if the vast majority of people think that expected utility theory gives the wrong results in decision-making, then, presumably, without any additional reason to accept the theory, it is in fact wrong.

There are thus serious questions regarding the strength of the basis of expected utility theory, and so serious questions about the plausibility of expected utility theory as the basis for a theory of reasoning under uncertainty. Still, this field is a flourishing one, furnishing, as it does, foundations for much of economics.

3.7 Other Arguments Equating Degrees of Belief and Probability

There are other, more difficult, theories for equating probabilities and degrees of belief. For example, an argument by Cox (1946 and 1961) places certain constraints on beliefs, which then turn out to obey the probability calculus. Another way of putting it is that Cox assumes beliefs can be measured. Then, given certain very weak constraints, this measure can be transformed into a probability. His argument is mathematically complex, and only now seems to be making some impact on the philosophical literature (although see Howson and Urbach 2006, Howson 2009, and Colyvan 2004). This could be because Cox required degrees of belief to be second-order differentiable: a somewhat curious restriction. Paris 1994 removes this restriction (but replaces it with another that is not intuitively obvious). Good references not only for this argument but for others can be found in Paris 1994 (which also contains a clear exposition) and Howson and Urbach 2006.

Finally, we should mention that recently James Joyce 1998 (not the Irish author) has produced an argument for the use of subjective probabilities based on the notion of estimating truth values. This idea is based on a generalization of the epistemological claim that (roughly) we should try to believe as many truths, and as few falsehoods, as possible. He argues that the notion of subjective probability gives a suitable generalization of this idea to partial beliefs, employing scoring rules (Appendix A.6.2), a generalization of Dutch Book arguments.

3.8 Is Bayesianism Too Subjective?

There are many complaints about the Bayesian approach's model of science, and in an introductory text they are far too numerous to survey. These complaints, however, mostly boil down to the claim that Bayesianism is too subjective. Consider Prokop's assignments in the solution to the Duhem-Quine problem. Bayesianism doesn't tell us why he chose the numbers he did, and so leaves open the possibility that they are completely arbitrary. If they are completely arbitrary, then Bayesianism doesn't actually represent a general solution to the Duhem-Quine problem: it only solves it for certain numbers. But why these numbers, rather than others?

Prokop might respond that they are not arbitrary at all: they reflect his experience, and his careful reflection on the evidence he has for and against his views. This is indeed true, but according to the Bayesian view, someone else with the same evidence might assign different probabilities. There is no way the Bayesian can rule this out. Indeed, that is why we use the adjectives 'subjective' and 'personalist' to describe this type of probability. One could argue that this means that the probabilities are in fact actually subjective, a reflection of taste, and not of anything objective. But science (and learning in general) is concerned with how things actually are, that is, with what is *objective*. Bayesianism, therefore, the opponent may conclude, cannot be an adequate description of science, for it is too subjective.

3.8.1 Bayesian learning theory

There are a number of responses to the objection that Bayesianism is too subjective.

The first step in any adequate response is to lay out the Bayesian account of learning from experience. This involves a discussion of how the probability of a hypothesis changes upon discoveries of evidence for or against it. We have already discussed the usual account of Bayesian learning, which characterizes learning as conditionalization, that is, upon the learning of e, $p(h)$ takes a new value, $p'(h)$, often conveniently calculated using Bayes's rule:

$$p(h|e) = \frac{p(e|h)p(h)}{p(e)} = p'(h).$$

The setting of $p(h|e)$ to $p'(h)$ upon learning e is called *Bayesian conditionalization*, where the value of h is conditional on e. (We used this principle implicitly throughout our discussion of the Duhem-Quine problem.)

The study of the behaviour of Bayesian conditionalization shows how the probability function behaves as more and more evidence accumulates. It turns out, perhaps not surprisingly, that given certain general conditions, the probability function becomes concentrated around a particular value depending on the evidence. If we are tossing a coin, and believe the tosses do not causally influence each other, and are carried out under the same circumstances, the value of the probability function will converge to the mean (to the proportion of observed tosses of heads/tails to total tosses). This is nothing other than an application of laws of large numbers, of which there are many different kinds.

3.8.2 Convergence of opinion

A particular application to the problem of subjectivism are 'convergence of opinion' results, which show that, under a broad range of circumstances, agents who begin with very different opinions but who are presented with the same information will gradually come to hold the same opinions. The most famous of these is probably the 'Principle of Stable Estimation', first introduced in Edwards, Lindman, and Savage 1963. I present a modification of their example.

The first graph (top left) in Figure 3.3 represents two probability distributions concerning the bias of a binomial parameter. As an example, let's use the belief of two scientists that a given coin is biased. The simplest way to represent this is by using beta distributions (the reasons for using beta distributions and the ways in which they are computed are not relevant to the point being made: please just enjoy the nice—and painfully produced—pictures). The first graph represents the prior beliefs of our two scientists that the coin will land heads. The distribution on the left, let's call it 'Prokop's', holds that the coin is most likely biased towards tails. The distribution on the right, let's call it 'Jarda's', holds that the coin is most likely biased towards tails. Not that Jarda is somewhat more biased than Prokop. The other graphs in Figure 3.3 show how the results of successive experiments—lowly coin tosses—change our scientists' prior beliefs. After 100 tosses their beliefs have fallen into line with the data. While Prokop and Jarda differed radically at first, they have quickly come to agreement. Most importantly, this agreement leads them far from their original beliefs, and is entirely data-driven.

This seems to answer the complaint that subjectivism is too subjective, given the dependence on prior probabilities. If prior probabilities did completely determine our views, then our views might only be a reflection of our biases. A standard claim is that convergence results show that biases fade in the face of evidence, at least for the Bayesian thinker.

It seems that this is a great success: Bayesianism leads us to the truth. This, however, is too much to claim. Consider a 'dogmatic' thinker, say, Vladimír again. He always thinks he is right, absolutely right, and so always assigns probability 1 (or 0) to all of his beliefs. The only way Vladimír learns is by falsification, but in no other way. It can be shown that, if he is not already correct, he will never become correct: once a hypothesis has probability 1 or 0, it cannot change—at least, not without a revolution in thought.

The convergence of opinion results relies on mathematical theorems of measure theory, and these are results 'excepting measure 0'. For us this means that the results hold if we are dogmatic in the right way: if we assign 0s where they are deserved, to events that really are impossible. The reader might wonder why we don't just avoid assigning any 1s or 0s at all, but some median value, and so be open-minded, unlike Vladimír. If we keep countable additivity (see A.2.4.1 and A.6.1), this is not an option, as there are too many propositions to assign probabilities to them all. Hájek 2003 contains an accessible discussion.

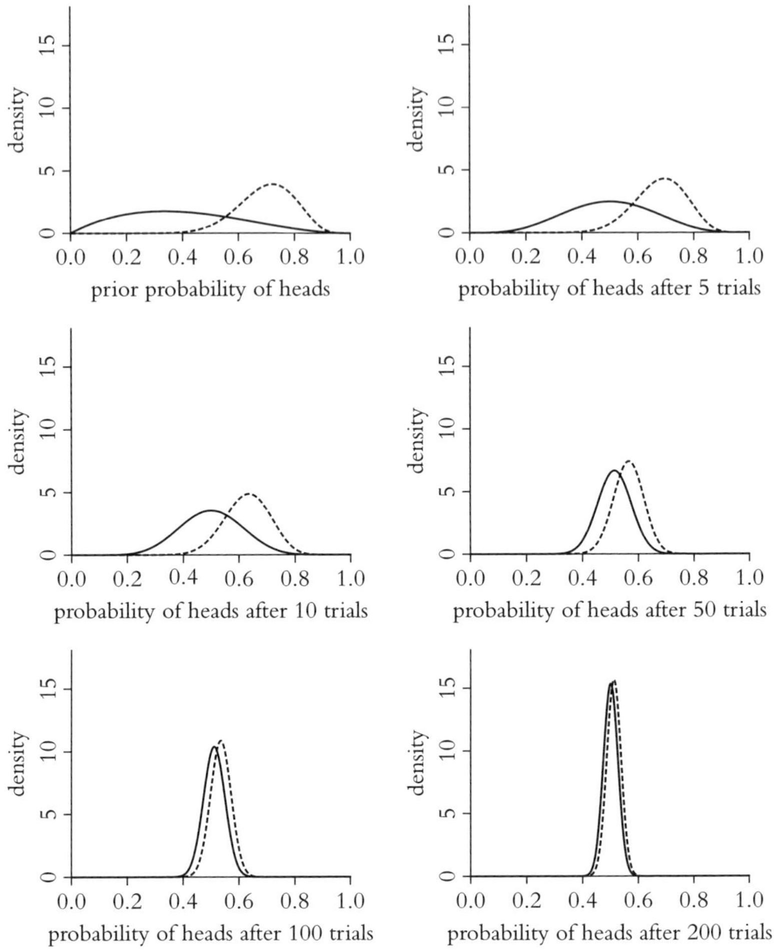

Figure 3.3 Convergence of opinions

The difficulty in the preceding paragraph was with assigning prior probabilities; another way of looking at it is in terms of the assignment of likelihoods. If we get the likelihoods wrong, evidence will not count towards the correct hypothesis, and we will not learn. Examples are, alas, rife. For example, suppose I believe in a very powerful god who created the world about 10,000 years ago. But, unlike Descartes's God, mine is a deceiver; he's gone out of his way to make the world look exactly like it would be if modern science is correct, i.e. very old, with life subject to the forces of evolution. Every bit of evidence you might bring against me either leaves my beliefs unaffected, or reinforces them. Now, it becomes rather difficult to maintain such views over the long run, but it doesn't mean that you can't. (If you don't believe me, have a look at explanations about how all earthly animal life might just have fit on a boat built using Bronze-Age technology, with room left over for 40 days' supply of food for them all.) A more frightening possibility: there really is such a god, and modern biology is badly off track.

Finally, Nature could be vicious, and attempt to hide any truths from us. For example, we might think that a number of coins are fair, having tossed them hundreds of times, and always getting a frequency of 1/2. But this could be the misleading initial segment of a sequence that has relative frequency of heads of, say, 3/4. If Nature is persistent enough, She can ensure that we will never converge on the truth.[4]

3.8.3 The problem of induction

All of these problems with convergence illustrate the general problem of induction. Observations of past phenomena can never yield certainty about the results of future observations. This was perhaps put forward most clearly and devastatingly by David Hume, and he seems to have had the last word. We can never get certainty for our theories from our observations (or anywhere else, Hume would have added). But the situation is even worse: no matter what observations we have made, any probability will be consistent with those observations given some background assumptions, which are not again forced on us by experience.

But we must employ some form of induction: indeed we often do. I do not think that the floor will collapse underneath me as I walk across it (again and again, pacing), I do not expect cola to come from Prokop's fermentation vats, and I do not expect my coins to start behaving very strangely. But why not? What makes us think that the world will behave in some regular fashion, so that we can draw on past experience to determine what will happen in the future? Presumably, we must find some principle that will allow us to make a justified leap on the basis of partial information to predictions, or at least confirmations of our beliefs about coins and

[4] Nature need not even be nasty: She can just be fiendishly complicated. Diaconis and Freedman 1986 examine cases where convergence fails; the same volume contains a discussion of their findings from a number of angles.

brains. But what is the status of this principle? It is not a priori, for there are possible worlds where no such principle applies. Indeed, for all we know, we may be in such a world. But if it is then a posteriori, we need to justify the principle. But the only principle that could justify the principle is the principle itself, which would be circular. (This version of the problem is from Kemp 2006: 5–6.) Ever since Hume, who famously raised the problem of induction, philosophers have attempted to circumvent it. It is uncontroversial to say that none have been successful. However, philosophers do divide on whether there might be some form of a solution.

To sum up, given certain assumptions, if we follow Bayesian rules we will converge upon the truth. If those conditions do not hold, we might not. For some this is a damning indictment of Bayesianism. For some Bayesians, it is the most reasonable way to approach an insoluble problem (or rather: the best way to approach a problem that has been solved—negatively). Other Bayesians argue that there are plausible constraints on prior probabilities that can to some degree ameliorate the problem of induction. This will be discussed in the next sections and chapters.

3.8.4 *Diachronic Dutch Books*

There is one further problem with the Bayesian account which aggravates the problem of induction: this has to do with conditionalization itself. The centrepiece of Bayesian methodology, and of the account of learning from experience, is that we change our degrees of belief by changing probabilities. This means that if we get some evidence e, we reset our probabilities to reflect this, setting $p(h|e) = p'(h)$, where p' is the new probability function. There is even a Dutch Book argument to back this up, due to David Lewis, but first reported by Teller 1973. (Lewis 1999 is a reprint of the original argument.)

Suppose we bet against $h|e$ (where $h|e$ means, as before, that the bet goes forward only if e occurs, otherwise it is cancelled), and on e. Now suppose that we bet on h after learning that e or not-e at a different rate. If e has occurred, the pay-off tables are

e	Pay-off
T	$+a$
F	$-b$

$h|e$	Pay-off
T	$-c$
F	$+d$

h	Pay-off
T	$+f$
F	$-g$

As can easily be seen, if $c > a + f$ and $g > b + d$, a sure loss is guaranteed. This can easily be achieved by changing the size of the stake. Therefore, the argument goes, we should condition our degree of belief in h; otherwise we are vulnerable to a Dutch Book. This is a *diachronic* Dutch Book, since it is concerned with bets entered into at different times. Previously the only Dutch Books we have discussed are *synchronic*, concerning bets entered into at the same time. (Naturally, the argument we gave in section 3.2.4 for the fourth axiom can be converted to a diachronic Dutch Book argument.) Supporters of this argument argue that Bayesianism is even more powerful than we thought. Not only does it impose a requirement of consistency for a given set of beliefs, it imposes a requirement of consistency of beliefs over time.

This claim has been hotly disputed. On the one side are objections that changes in degrees of belief need not be via conditionalization, but by some other reasonable means. For example, your degree of belief may change not because of any particular evidence, but, perhaps, because of a broader change in your philosophical outlook. Perhaps you have mellowed with age, and no longer so violently hold beliefs. Or perhaps the opposite: you rage with the nearness of death. These changes do not seem to be irrational. Yet, if the Dutch Book argument for conditionalization is accepted, they must be so classified. On the other side, some find it difficult to understand how an ideally rational bettor would change his or her beliefs other than by acquiring new evidence, in some form or other. We will not go into this debate, but refer the reader to the classic papers of van Fraassen 1984 and Christensen 1991.

The standard defence of conditionalization is to hold that the principle applies given certain restrictions: namely, that there has been no change which 'disturbs' the probabilities. While this offers an interesting illumination of the limits of Bayesian learning theory, it does not, of course, offer a general solution to the problem of induction.

3.9 Is Bayesianism Too Flexible? Or Not Flexible Enough?

One of the reasons that many hold a Bayesian account of learning is that it has been said to be very successful. We covered one of these success stories earlier. But we also saw that the success is dependent on the application of certain priors. If the priors were different, then there would be no success. It may be that Bayesianism accords with our intuitions when we put in the right priors, but this may also be an artefact of the great flexibility of

Bayesianism. Given the right priors, it can explain anything. This claim is part of the reason Bayesianism is criticized for being too subjective: without objective grounding, Bayesianism can explain everything, and thus, perhaps, nothing. One possible, quite Bayesian, response would be that the probability calculus (interpreted as the calculus of degrees of belief) is a tool, and its adequacy as a tool relies on its capacity for modelling our intuitions correctly, and providing a means for ordering them. This will be far too liberal for those who believe that there is a means for somehow directly accessing truth. I have no stand to take on this debate, being constitutionally sceptical.

Another worry about Bayesianism is that it leaves out certain important, perhaps qualitative, features of good scientific theories. For example, traditionally, good theories have been held to be able to explain new and old facts in an illuminating way, to be simple, to fit with other theories having these characteristics, and so on. A recent trend in the philosophy of science, known as scientific realism, holds that our best theories are (approximately) true, and possess certain of the traditional virtues. But, as Milne points out, Bayesianism only gives an account of the weight of evidence for or against a hypothesis given other weights—not of the quality of the evidence as such. Additional considerations like theoretical virtues therefore not only fall outside of the Bayesian framework; they conflict with it, unless they reduce to some sort of weighting in, say, the setting of prior probabilities. But such weighting would not be what a scientific realist would desire of theoretical virtues, since the weights are, ultimately, subjective attitudes to evidence. For a discussion, see Milne 2003. For a discussion of the criticism that Bayesianism does not distinguish between novel predictions and ad hoc adjustments in a theory to account for predictions, or that it can't even account for the use of evidence already known, see Howson and Urbach 1993, sections 15.g and 15.h.

We could also question the putative link between truth and the non-Bayesian virtues. Presumably, there must be some non-Bayesian inductive account providing such a link. But none seems to be forthcoming. Indeed, the link between an ontological notion (laws of nature) and epistemic and aesthetic notions (simplicity, beauty, illumination, etc.) is an odd one: for things are true no matter what we may think of them. One way around this would be to employ a pragmatic notion of truth, but we would then no longer be dealing with realism. And, of course, even for a theory of pragmatic truth we would need an account of the link between evidence and truth.

3.10 Conclusion

I have tried to give an overview of the main arguments for equating degrees of belief and probabilities, as well as of the objections to these arguments. It is not possible to survey them all; the literature is too large. Hopefully the reader will be left with the impression that there are interesting arguments for subjective probability, about which there is illuminating controversy. I have not drawn parallels with what amounts to the same controversies in other parts of philosophy, but the reader should note that wherever there are questions about knowledge and belief, the same issues arise.

I have already alluded to the large literature on Bayesian thought. This includes a number of introductory books and articles. Foremost among these are Howson and Urbach's *Scientific Reasoning*, my preferred edition being the second (Howson and Urbach 1993). As well, the introductory texts on the philosophy of probability by Gillies 2000, Mellor 2005, Galavotti 2005, and Hacking 2001 contain differing perspectives on Bayesian thought. Earman 1992 is a noteworthy earlier contribution.

4

Subjective and Objective Probabilities

In the preceding chapter we looked at subjective probabilities as grounding a theory of inference. This leads naturally to the question of the relation between subjective and objective probabilities. One standard position is dualism: there are both objective and subjective probabilities, and they can be combined by letting values of objective probabilities serve as evidence in Bayesian calculations. This dualistic approach fits nicely with the view that objective probabilities are features of the world uncovered by science, and that Bayesianism provides a theory of inference for science. The bonus, for some, is that Bayesianism becomes more objective. Justifying dualism from this point of view would seem to require Dutch Book arguments. Apart from its justification, dualism also leads, when combined with a particular kind of empiricism, to deep metaphysical problems. In opposition to the dualist view, de Finetti held a monist view according to which what are taken to be objective probabilities are simply a species of subjective probability. This chapter will cover these issues in order.

4.1 The Principle of Direct Inference

Many Bayesians hold a dualist account of objective and subjective probabilities. For example, it is natural to think of a sequence of coin tosses as a series of independent and identical trials with the same probability of outcomes, where the probability of the coins coming up heads is taken to be objective (in the jargon, the outcomes of the tosses can be represented by independent identically distributed, i.i.d., random variables). A subjective probability distribution can then be defined over the possible values of the objective probabilities. So, for example, we can take the bias of a coin as an unknown parameter, and use the outcomes of coin tosses to attempt to determine the value of the parameter. This is essentially the same as the convergence of opinion in section 3.8.2: data draws the subjective probabilities closer to the true value. (As we shall see in section 4.4, strict Bayesians firmly reject this interpretation: that's why they call it convergence of *opinion*, and not convergence to the truth.)

The question of how to include frequency data in Bayesian calculations has traditionally been framed as a question of direct and indirect inference. The earliest accounts of probability concerned reasoning from knowledge of the frequency of an attribute in a population to the frequency of that attribute in a sample. The traditional name for this reasoning is direct inference, while Bayesianism, reasoning as it does from samples to populations, from evidence to hypothesis, has traditionally been named inverse inference. (The emergence of the term 'Bayesian' is explored in Fienberg 2006.)[1]

Let h be the hypothesis that the coin has bias x (.5 meaning the coin is fair, 1 that the coin is completely biased to heads), and H that the coin comes up heads. Then $p(\mathrm{H} \mid h = x)$ is the direct probability, and $p(h = x \mid \mathrm{H})$ the inverse probability (recall their current names, happily or not, are 'likelihood' and 'posterior probability'). The question is then, what values should we assign to direct probabilities? Answers to this question are Principles of Direct Inference, a wide variety of which are on offer (Kyburg 1981: 773 lists a number of proposals). A simple, and seemingly blindingly obvious Principle of Direct Inference would be to require that $p(\mathrm{H} \mid h = x) = .5$. But no contradiction arises from setting $p(\mathrm{H} \mid h = x)$ to something other than .5, and so it seems we need a justification for such a Principle.

Hence the search for an argument that knowledge of frequencies should constrain our beliefs about frequencies in a sample. The usual way of producing a justification for constraints on beliefs, as we have seen, is by providing a Dutch Book argument. The next section is thus devoted to such arguments.

[1] In this context it is worthwhile noting that the distinction between direct and inverse inference is fundamental in statistics. Current statistical practice is dominated by theories of direct inference (usually called frequentist statistics, although sometimes, and somewhat oddly, called classical statistics). While we will not discuss frequentist theories of inference (for which see Mayo 1996), it is worth stressing that frequentist interpretations of probability need not yield frequentist theories of inference. The two problems, interpretation and inference, are separate, a point often overlooked, which is perhaps why, against all the evidence, von Mises gets labelled a frequentist in both aspects.

4.2 Betting on Frequencies

How, then, should our subjective probabilities be related to frequencies generally, and single-case probabilities, chances, in particular? One obvious idea is to think of betting on mass phenomenon (and for von Mises as well, given the motivation of the axiom of randomness by the notion of the impossibility of a gambling system). Suppose you are observing a series of coin flips, which you regard as part of a collective, and, for whatever reason, wonder what you should assign as a probability to the next coin flip coming up. Howson and Urbach 1993, on the basis of a Dutch Book argument, advise you to assign the value equal to the limiting relative frequency of the attribute (heads in this case) in the collective. The argument is as follows: suppose you are certain that the next toss will be a member of a collective with limiting relative frequency r and that you offer a fair betting quotient p different from r. Then, in a series of bets (that is, subjective probabilities of a series of trials), one side will eventually be subject to a sure loss. (Suppose the limiting relative frequency of heads is .5, and you offer a betting quotient of heads of .7. Then you will in the limit lose 20 per cent of the time.)

Howson and Urbach claim that this gives empirical content to von Mises's theory, since we've established, using Bayes's theorem, the likelihood that we are dealing with a particular collective. The value we need to determine is $p(C \mid A_i = x)$, where C is 'this series forms a collective with limiting relative frequency r' and $A_i = x$ is 'the value of the ith member of the collective is x', namely, the one under consideration right now. To use Bayes's theorem, we need to establish $p(A_i = x \mid C)$. If Howson and Urbach's argument is correct, we have established that it is equal to r. Some elementary probability considerations (found in Howson and Urbach 1993: 344–7) show that in a series of tosses, the subjective probability will very quickly converge to the true value, r. (Mellor 1971: 163 offers a similar argument, but for his account of propensities, not collectives.)

It might seem that Howson and Urbach have proved too much and provided a solution to the problem of induction. For we have now shown, it seems, that our subjective probabilities must be equal to the limiting relative frequency. Not so—our personal probabilities are only determined by being certain that we are dealing with a collective, and furthermore, that the collective has the limiting relative frequency we think it does. Alas, we can be wrong about both of these (as discussed in sections 3.8.2 and 3.8.3). Albert 2005 takes this failure to establish a link between degrees of belief and the true relative frequency to show that Howson and Urbach's attempt to give empirical content to von Mises's theory fails. This is not entirely correct, since Howson and Urbach's aim is not to solve the problem of induction, but to show that under certain circumstances hypotheses about collectives can have empirical import. In this they seem to have succeeded. (For other efforts in this direction, see Romeijn 2005; Howson 2000: 207–8 responds to Albert's argument in much more detail.)

There is, however, a rather more serious problem: the Dutch Book does not establish that each individual bet, that is, each individual subjective probability, should equal the relative frequency. Suppose you are to bet on the toss of a fair coin that you believe would form part of a collective. What should you bet? If you are betting on only one toss, it's hard to see that you are committing yourself to a sure loss if you bet at, say, 3 to 1 odds. Maybe you're willing to take the extra risk, because you feel lucky. And you might be: you might get a pay-off of 3. This holds for any finite series of bets: nothing in probability theory shows that you are certain to lose if you bet against the relative frequency finitely many times. In fact, probability shows exactly the opposite: deviations from the mean are to be expected for finite cases. So there is no *certain* loss. But this is necessary to make the Dutch Book argument work, for the bettor has to be indifferent between a sure loss and a sure gain. But the bettor is not so indifferent in this case.

Howson and Urbach's argument rests on moving from a collection of bets to each individual bet: if we consider a collection of bets together fair we should consider each individual bet fair. This is not obvious, as we have just seen. In fact, Strevens (1999: 262) argues that this assumption is equivalent to assuming a Principle of Direct Inference. (Note that the move in the opposite direction is not so questionable: if we regard a series of bets as individually fair we should regard the set of bets fair as well.) It therefore seems that bets, hence betting quotients, hence degrees of belief, are unconstrained by objective probabilities.

Childers 2012 contains more reflections on Dutch Book arguments for Principles of Direct Inference; Pettigrew 2012 applies Joyce's non-pragmatic justification mentioned in 3.7 to a justification of a version of a Principle of Direct Inference discussed in the next section.

4.3 The Principal Principle

One particular Principle of Direct Inference has come to dominate the literature—David Lewis's Principal Principle. (While there are surely many more principal principles than this one, there are also surely more direct inferences than those covered by the Principle of Direct Inference.) One formulation Lewis gives of the Principle is:

$$ch_{tw}(A) = p(A \mid H_{tw} T_w)$$

where *ch* is an objective probability, or, as Lewis refers to it, chance, *p* is a subjective probability, *A* a proposition, *t* and *w* indexes that range over times and worlds respectively, T_w a complete theory of chance for world *w* and H_{tw} the history of world *w* up to time *t*. Chance is in this case taken to be objective and apply to single cases. Thus stated, the Principal Principle is obviously compatible with a propensity interpretation like that outlined in 2.2.6, as Lewis meant it to be. A 'complete theory of chance' is 'a full specification, for world *w*, of the way chances at any time depend on history up to that time' (Lewis 1980: 97). In particular, Lewis conceives of the complete theory of chance of being composed of what he calls 'history to chance conditionals', which are, unsurprisingly, conditionals with histories as antecedents and values of chances as consequents. The complete theory of chance is constructed to exclude what Lewis calls 'inadmissible' evidence, that is, the 'forbidden subject matter—how the chance processes turned out' (Lewis 1980: 96). This applies, for Lewis, to statements about the future concerning the outcome, since he takes the past to no longer be chancy: the coin was tossed, and it landed either heads or tails, and so there is no longer chance involved. So the complete theory contains no precognition, prophecies, and the like (concerning outcomes under consideration). Admissible evidence is usually characterized along the lines that it changes probabilities of outcomes only in terms of the chances of the outcomes, and not the outcomes themselves.

4.3.1 Humean supervenience and best systems analyses of laws

One reason for the interest in Lewis's Principle is its seeming incompatibility with his broader programme of Humean supervenience. Humean supervenience is the view that 'the whole truth about a world like ours supervenes on the spatiotemporal distribution of local qualities' (Lewis 1994: 473).

> [Humean supervenience] says that in a world like ours, the fundamental relations are exactly the spatiotemporal relations: distance relations, both spacelike and timelike, and perhaps also occupancy relations between point-sized things and spacetime points. And it says that in a world like ours, the fundamental properties

are local qualities: perfectly natural intrinsic properties of points, or of point-sized occupants of points. Therefore it says that all else supervenes on the spatiotemporal arrangement of local qualities throughout all of history, past and present and future.

(Lewis 1994: 473)

The impulse behind this programme is empiricist: there is a fundamental stock of things in the world (say, point-masses), and everything else is to be accounted in terms of relations, patterns of the fundamental things. Laws, causation, and other problematic (from the empiricist point of view) relations are accounted for as patterns in which these objects may be arranged. Hence 'Humean' for its empirical sparseness, 'supervenience' since items like causation supervene on the patterns, in that any change in a supervening item means a change in the underlying patterns of fundamental things.

Chance must also supervene on the world, in terms of being an arrangement of 'local' qualities. The obvious way to explain chance in terms of patterns in the world would be to use relative frequencies. Lewis rejects the use of relative frequencies to explain chance for a number of reasons (his example, discussed in 2.1.2, of a seemingly non-mass objective probability attaching to radioactive decay; or that frequency accounts don't assign probabilities to unrealized situations, see 2.1.1). Lewis's answer to how propensities can supervene on local arrangements is to retain a dispositional, single-case account of probability, but appeal to laws as determining the values of the disposition. This, of course, is no answer until we have an account of laws compatible with Humean supervenience.

Lewis's own favoured account of laws compatible with Humean supervenience is a 'best systems analysis':

Take all deductive systems whose theorems are true. Some are simpler, better systematized than others. Some are stronger, more informative, than others. These virtues compete: an uninformative system can be very simple, an unsystematized compendium of miscellaneous information can be very informative. The best system is the one that strikes as good a balance as truth will allow between simplicity and strength. How good a balance that is will depend on how kind nature is. A regularity is a law iff it is a theorem of the best system.

(Lewis 1994: 478)

The situation is different if we include probabilistic laws. These laws are chosen for their best fit to series of events under consideration: but because they are probabilistic, the fit will not be perfect. Therefore laws can be very simple, but not completely accurate, allowing for many exceptions; they can also be strong, allowing for few exceptions, but this usually results in greatly increased complexity.

We could account for the law of radioactive decay as a summary, more or less accurate, of occurrences of decay (that is, it matches up nearly enough with observed frequency). But a frequency reading is not necessary: in cases where there is no mass phenomenon the probabilistic law will not be determined by frequency, but by, say, symmetry considerations, or how well the law fits with the other laws.

4.3.2 The big bad bug and the New Principle

We are now in a position to see the conflict of the Principal Principle with (Lewis's version of) Humean supervenience. The best systems analysis of laws shows us how chance can supervene on non-chance properties: the values of chances are given by the best system. However, this also leads to a contradiction. Lewis conceives of undermining futures, futures which, if they come about, make the *present* chances different.

He gives an example of a radioactive element (tritium) having a different half-life. Remember that laws of radioactive decay in Lewis's Humean account supervene on local matters of fact, and so in this particular case on how tritium decays. So it can't be ruled out, right now, that tritium will decay differently in the future (although given how it has behaved up to now, the probability that it will decay differently is very small—no matter, it is still non-zero). Put differently: there are no prior laws from which the Humean can derive how tritium behaves; the laws are, on balance, (best systems) summaries of how tritium actually behaves. The future being open, it may behave differently, so there is a non-zero chance that this odd future will come about.

For illustrative purposes, take chance to be actual frequency. (We will use this to generate a 'toy example'; more complicated versions could be concocted for other best systems analyses of chance; see Lewis 1994: 488.) An actual frequency account gives us a simple complete theory of chance, which, with history, gives us our chances. We have a coin, which we have tossed 10 times, and it's come up heads 5 times. So, the complete theory of chance in conjunction with history entails that $ch(\mathrm{H}) = .5$. We can calculate the chance of the coin coming up heads each time for the next 100 tosses, after which it will be destroyed. (We use the binomial theorem, see A.5, for 95 tosses with 95 heads, with $ch(\mathrm{H}) = .5$.) This has a very small probability indeed, .00000000000000000000000000002524 . . . That's 28 zeros in all, i.e. 2.524×10^{-29}.

But, if this future comes about, the relative frequency of heads, and hence the chance of heads, will be 95/100. This does not mean that in the future the relative frequency will be .95—it means that the frequency right now will be 95/100. Applying the Principal Principle, then, $p(95 \text{ heads out of } 100 \mid H_{tw}T_w) = 2.524 \times 10^{-29}$. However, 95 heads out of 100 implies, again, that the chance of heads is .95, which contradicts the conjunction of history up to now and the complete theory of chance, which implies that the chance is .5. Hence, the subjective probability of 95 heads coming up must be zero, $p(95 \text{ heads out of } 100 \mid H_{tw}T_w) = 0$. We therefore have two values for the subjective probability, and a contradiction.

To put it differently: the possibility of a different objective probability in the future contradicts our *present* theory of chance, T_w. After all, the undermining future is undermining just because it contradicts present chances. Our theory of chance on which we condition is determined only up to now, and so excludes the possibility of undermining. Hence the subjective probability for this should be zero. But by the Principal Principle it should be the same as the objective chance probability, which is non-zero. Hence a contradiction, and a seeming death blow to any attempt to account for chance in a strictly Humean framework. We seem to be in need of a way to pin down objective probabilities in a way that doesn't seem available within Lewis's framework. Lewis calls this contradiction the 'big bad bug'. (Lewis 1986b: xiv.)

This difficulty led (in the trio of papers Thau 1994, Hall 1994, and Lewis 1994) to the 'New Principle':

$$ch_{tw}(A \mid T_w) = p(A \mid H_{tw}T_w)$$

Instead of unconditional chances, the values are now conditioned by our complete theory of chance. Chance is no longer absolute, but conditional on the complete theory of chance, which will exclude the undermining futures. In our example above, $ch(95 \text{ heads out of } 100 \mid T_w) = 0$, since our theory of chance (right now) sets the chance at .5, and 95 heads coming up sets the chance at .95, a contradiction. Hence, conditioning on the current theory of chance frees the Principle from contradiction (the chances being constrained on both sides of the equations).

The New Principle also represents a rather drastic limitation on probability assignments. In a small sample such as ours, the probability of getting any number other than 50 heads in the 100 tosses has probability 0. The charts in Appendix A.5.3 show just how much this differs from the usual way of assigning such probabilities. For example, in 30 tosses the chance of getting exactly 15 heads is very low, while, of course, the chance of getting nearly 15 heads is quite high. Lewis (1994: 488) points out that when dealing with small samples from large populations this effect is very small. For example, the chance of three heads after 100 tosses out of 1 million will be very near the same using the usual formulation of A.5 and Lewis's approach. Still, the Bayesian approach is often thought of as one that can be used in small samples, and so the dramatic difference is worth noting.

A large literature has sprung up around these issues, and whether the modification is necessary, or sufficient, to avoid contradiction. Briggs 2009a and 2009b provide a sceptical overview of various combinations of Humean supervenience and objective probabilities. Carl Hoefer 2007 makes a notable contribution with his Humean framework which allows 'deterministic chances'.

An introduction to Lewis's views on Humean supervenience can be found in chapter 4 of Nolan 2005. Best systems analyses have begun to flourish: see for example Callender and Cohen 2010. Since these theories must take into account probabilistic laws, they are intimately connected with the difficulties discussed in this section.

We have come across yet another case of a question of the interpretation of probability leading out to deep metaphysical waters. We still lack, however, a justification for a Principle of Direct Inference. Lewis held that the Principal Principle should be taken as summing up what we know about chance. Since it serves as an adequate analysis of the concept of chance, no further justification is needed. The strength of this claim will depend on your views of philosophical analysis.

4.4 Exchangeability

De Finetti was a monist regarding probability: he held that there is only one correct interpretation of the probability calculus—the subjective interpretation. But he was well aware that there is much objective-sounding talk regarding probabilities of the sort in the preceding paragraph. He therefore set himself the task of showing how to translate this talk into talk about subjective probabilities alone, thus allowing for an elimination of objective probabilities—a fine example of an early twentieth-century philosophical project. If the project were to succeed, it would make the rest of the accounts covered in this chapter at best otiose.

De Finetti's account turns on the notion of exchangeability. We will consider the simplest case, which is, as always, the coin tossed indefinitely. We call an assignment of probability to binomial random variables representing coin tosses (finitely) exchangeable if the probability does not change when we permute the order in which we consider them. So, if we are considering a sequence of three tosses of a coin resulting in two heads, we consider the sequences HHT, HTH, and THH to have the same probability (which implies that TTH, THT, and HTT have the same probability as well). It is important to keep in mind that this is a *subjective* assessment of the chance of two heads in a row. A stronger condition is exchangeability proper: this is a property of probability assignments to infinite sequences of binomial random variables, namely that any finite subsequence of random variables is exchangeable. De Finetti showed that an exchangeable sequence of random variables is, formally, the same as a subjective distribution over unknown objective probabilities (the resulting probability distributions are the same). That is, for any subjective probability distribution over i.i.d. random variables with constant objective probability there is a corresponding solely subjective probability distribution over exchangeable random variables.

This provides a reduction, via a representation, of physical probabilities, at least in this case. The i.i.d. random variables are meant to represent independent trials under constant conditions. But exchangeability allows us to replace the notion of independent conditions with a restriction on degrees of belief concerning probabilities, namely, exchangeability. Put differently, we can substitute expectations of a certain kind (exchangeability) for postulating objective probabilities (the probabilities associated with i.i.d. trials).[2]

A concrete example: Prokop has a trick coin, given to him by his charming rogue of an uncle, Pavel. It's a trick coin because it's very heavily weighted, two to one, towards one side. The problem is, Prokop can't remember which side. So he decides to generate some data, tossing the

[2] De Finetti gave his most influential account of exchangeability in 1937. Heath and Sudderth 1976 give an accessible introduction and proof of the convergence of exchangeable distributions. While we discuss only simple binomial trials, it should be noted that the notion of exchangeability can be generalized to random variables with more than just two outcomes. It may also be helpful, in light of Chapters 5 and 6, to think of exchangeability as a symmetry condition imposed by ignoring the order of the tosses.

coin. A dualist would say that Prokop is looking for the probability of the coin landing heads; a monist can respond that there is no need to postulate an underlying objective probability. Instead, Prokop holds that the trials can be represented by exchangeable random variables, each of which influences his degrees of belief that the coin lands heads. In this case he assigns probability 1/2 to the hypothesis that the coin is weighted 2/3's to heads, and 1/2 to the hypothesis that it's weighted 2/3's to tails. Thus he considers the sequences with 2/3 heads to be exchangeable, as those with 1/3 heads. These two sets of exchangeable sequences received probability 1/2, all the others probability 0. They thus determine a likelihood value: an observed heads supports the hypothesis that the coin is biased to heads, tails the converse. This procedure is formally the same as taking the bias to be an objective probability as in 4.1.

De Finetti gave another motivation for exchangeability: it is natural for a Bayesian to expect that his or her opinion will change in the course of a series of repeated trials. Therefore, degrees of belief at each stage will be dependent on what came before, and so independence is not a natural condition for the Bayesian. Instead, the Bayesian should expect that a series of outcomes given by an experiment would not be affected by the order of the outcomes, since the trials are undertaken separately of each other. (You might want to consider Prokop's trick coin in this context.) Exchangeability is a surrogate for objectivity, since it implies that the underlying physical process will in the end give the same results—from the point of view of the observer—no matter the particular way it is realized.

De Finetti therefore claimed to have found a way to account for the notion of an unknown objective probability. The importance of this claim is that all probabilities can now be shown to be epistemological, thus establishing a form of monism. Moreover, this is an advance in the epistemology of the sciences, since we now only have to account for subjective probabilities alone, not both objective and subjective. Still, we must not expect too much: for the problem is that we must be sure that the sequences *are* properly thought of as exchangeable, otherwise we will be misrepresenting the experiments, and may draw the wrong conclusions. This, again, is the problem of induction we met in 3.8.3, for any principle which would tell us when to apply the notion of exchangeability cannot be necessary—for there are worlds in which the principle would not work (say, one inhabited by a malicious demon who always makes sequences look exchangeable in the short run but ensures that they are not in the long run). On the other hand, we cannot determine the principle a posteriori, for we need the principle to establish all a posteriori truths, and to apply it in this way would be circular. (Again, my description of this version of the problem of induction is inspired by Kemp 2006: 5–6.) De Finetti would have probably agreed with this point, since he was a subjectivist: his did not aim to show how we could (infallibly) arrive at correct answers about physical processes giving rise to probabilities, but rather how we could eliminate the very notion of such processes.

One difficulty raised by Howson and Urbach (1993: 349–51), which they attribute to I.J. Good, is that the very generality of exchangeability limits its appeal in contrast to the assumption of physical probabilities. They argue that the assumption of physical probabilities is more appealing than that of exchangeability, no matter the possible gain in metaphysical austerity from a monistic interpretation of probability. We can use the following example drawn from Zabell's illuminating 1998 paper (11–12): suppose we have two sequences which we think are exchangeable, and which share the same number of successes, but which differ in their 'orderliness'. Specifically, suppose the 'orderly' sequence is one in which every H is followed by a T, while a T may be followed by either an H or a T. At a given trial the toss lands H. What is the probability that the next toss will be H? According to the assumption of exchangeability, it is equal to the relative frequency of H's in the sequence. But this seems a most peculiar assumption to stick with: it seems much more reasonable to drop the assumption of exchangeability and instead set the probability of the next toss being T higher. (De Finetti discusses this issue in 1937: 145–55.) Anyone making the assumption of exchangeability would seem to be relying on an implicit assumption that the tosses really *are* independent, in some physical sense —that the tosses don't influence each other. If you are convinced that they are so independent, then you can dismiss the oddly regular sequence as a fluke. But then it seems to make more sense to simply postulate physical probabilities instead of exchangeability. However, it seems clear that the proponent of exchangeability could simply retort that since independence implies exchangeability, but not vice versa, the odd attitude to take is that the trials are independent, since it carries more, and unnecessary, ontological baggage.

4.5 Conclusion

If none of the justifications for a Principle of Direct Inference presented in this chapter are correct, then subjective probabilities are truly subjective: they are free from the constraints of objective probabilities. If de Finetti has succeeded, we arrive at a quite radical subjectivism (of which he surely would have approved). Another way of looking at the problem (if it is one) is in terms of reference classes: Bayesianism does not face the reference class problem because the determination of what should be considered for inclusion into a series is treated as a subjective matter. But this doesn't seem to be much of a solution: presumably we would like to determine objective probabilities *objectively*. So, the combination of subjective and objective probabilities leads us back to the problem of the reference class. This problem is, of course, a problem of induction. But Bayesianism can only offer a very limited solution, as seen in section 3.8. Hájek 2007 argues that this creates great difficulties for Bayesianism (and for the other interpretations of probability as well).

Many people find the notion that we are doomed not to know the future annoying (or even a scandal). This perhaps explains the constant search for principles to add to the core Bayesian ones to gain some security in an uncertain world. Versions of the most popular candidate will be explored over the course of the next two chapters.

5

The Classical and Logical Interpretations

Prokop has the urge to share his new-found knowledge about probability with his friends. When they can concentrate, they all say pretty much the same thing: 'Uh huh. I'm thirsty'. Actually they usually say 'I always thought the chance of a coin landing heads is 1/2'. This deeply irritates Prokop, but he always patiently responds 'It depends. It depends on what you mean by chance. It depends on the coin. It depends what you believe about the coin'. He has discovered that if he attempts to elaborate further, his friends suddenly become overwhelmed by their thirst, and leave to rummage through his (nice, big, American) refrigerator. The idea that probability can always be determined by distributing it over different possibilities in a symmetric way is deeply entrenched. 'But why?' wonders Prokop.

5.1 The Origins of Probability—The Classical Theory

The interpretation of probability Prokop's friends allude to is the oldest. It originates in correspondence between Pascal and Fermat (although, of course, there are earlier precedents: many are surveyed in Franklin 2001). Pascal and Fermat dealt with a problem of how to distribute winnings when a fair game is interrupted before completion. (We won't discuss this problem; an account can be found in Franklin 2001: 306–13.) Their correspondence was well known (correspondence being the mode of transmission of mathematical results at that time), and inspired Christiaan Huygens to publish the first book(let) on probability, *De Ratiociniis in Ludo*

Aleae in 1657. (The interested reader can find an English translation in the impressive Verduin 2009.) The methods of probability were developed by many authors, the most significant contribution perhaps being Jacob Bernoulli's posthumously published *Ars Conjectandi* (Bernoulli 1713). The summing up of the classical interpretation, which was to dominate up to the twentieth century, was done by Pierre-Simon, Marquis de Laplace in his 1814 *Essai philosophique sur les probabilités* (*Philosophical Essay on Probabilities*).

The key to the classical interpretation is the notion of symmetry. The *Oxford Concise Dictionary* tells us that 'symmetry' means 'the quality of being made up of exactly similar parts facing each other or around an axis' or 'correct or pleasing proportion of parts'. In the case of classical probability, probabilities are assigned symmetrically (i.e. equally) to certain basic possibilities given there is no reason not to do so. The probability of an event is defined as a ratio of possibilities: those possibilities in which the event occurs divided by the total number of possibilities. This is usually put as: the probability of an event is the number of events *favourable* to that event divided by the whole number of events. We will follow Keynes 1921 in naming the principle that we should so assign probabilities the Principle of Indifference (it is sometimes also known as the Principle of Insufficient Reason).

The classical theory holds that probability measures ignorance (or partial knowledge), since probabilities are assigned because of lack of reasons to do otherwise. For this reason it can also be seen as a measure of the degree of rational belief (although, as we shall see, it's not that simple). Simple examples suffice to illustrate. Suppose we toss a coin about which we know nothing. There are only two possibilities, and so heads takes up 50 per cent of the 'possibility space'. So, we assign 1/2 to its coming up heads, and 1/2 to its coming up tails. The same goes for a roll of the die about which we know nothing: if it is the regular six-sided kind, the probability of the die landing with a particular face up is 1/6 since it takes up 1/6th of the possibility space. In general, given n options, and no reason to differentiate the options, the probability assigned to any one of the options is $1/n$. This function is, rather trivially, a probability.

Probability distributions created by the Principle of Indifference are in a sense *flat*, as we can see if we chart the amount of probability by the amount of possibilities. Figure 5.1 shows the probability of a fair coin over the options:

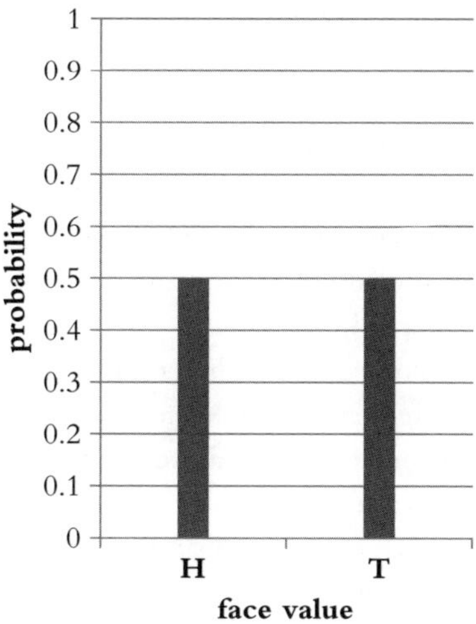

Figure 5.1 Coin tossing

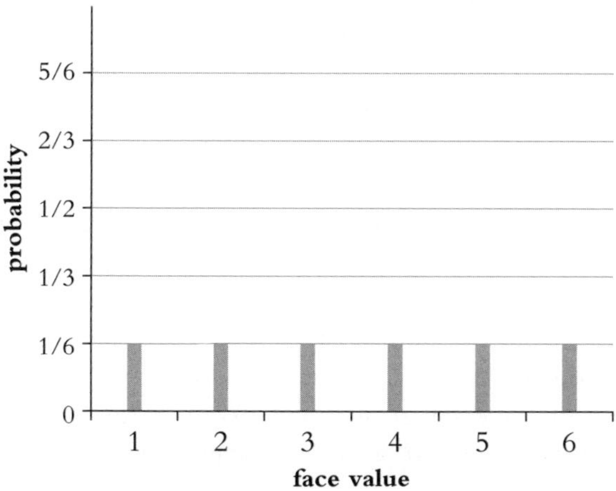

Figure 5.2 Rolling a die

Since the probability of each possibility is the same, when we graph, as usually, probability with respect to probability we get a flat line. Figure 5.2 shows what we get for a roll of the die.

As already noted, the interpretation was developed through considerations of gambling. This gives much more generality than might be expected: a vast amount of problems in probability can be expressed as gambling problems. Moreover, the classical interpretation is quite easy to apply. To give an example of how the classical interpretation works, consider the probability of getting 4 when rolling two dice. Each die has 6 possibilities, giving a total of 36 possible outcomes when rolling two dice. If we know nothing more, the classical interpretation tells us to assign probability 1/36 to each outcome. There are three combinations that give a combined total of 4: 2,2; 1,3; 3,1. The probability is therefore $3/36 = 1/12$. (Combinations are explored more in depth in A.4.)

Calculating the outcomes of our dice rolling is simpler than with the Bayesian and relative frequency approaches: we don't have to determine prior probabilities; we don't have to find a collective. We just need the sample space and ignorance and the determination of probabilities is simple. It thus serves as a model for determining probabilities for fair chance games and phenomena that can be modelled as such games.

5.1.1 The Rule of Succession

As should now be obvious, classical probability is very much involved with the study of combinations—combinatorics. There is a brief consideration of some of the basics of combinatorics in A.4. The use of combinatorics allows us to prove, among many other things, of course, a famous result. Assume that we are observing a sequence of independent events of which we do not know the probability. Further assume that we are, as a classical probabilist would put it, completely ignorant of the probability, and so consider all possible values of that probability equally likely. (From a Bayesian point of view we could say that we assign a flat subjective probability distribution to the objective probability of the event.) Then it can be proved that the probability that the event will repeat given n previous observations of the event (in the circumstances appropriate for the occurrence of the event) out of m observations (again, in the appropriate circumstances) is

$$\frac{n+1}{m+2}.$$

This is called, following Venn 1876, the *Rule of Succession*. The Rule is proved using an urn model, that is, by considering the drawing of balls from an urn containing white and black balls only, where we know nothing about the proportion of white balls. The Rule is derived by letting the number of balls in the urn grow arbitrarily large, and applying the Principle of Indifference. (We will not prove the Rule, since doing so is neither easy nor necessary for our present purposes—for a proof of a version of the Rule and much more, see Zabell 1989.)

The Rule is of surprising power: it tells us to expect that the probability of the next occurrence of the event (e.g. the drawing of a white ball, rain after a falling reading on the barometer, algae blooms after Saharan sandstorms) to be, for a large enough number of occurrences, near the relative frequency of previous occurrences. This seems to provide a solution to the problem of induction, since it gives a putatively *objective* probability of the occurrence of the event. (Although the Rule never gives us certainty: if the relative frequency is greater than 1/2, the rule gives us a slightly greater probability, and if it is less than 1/2, it gives us a slightly smaller probability. If the event never occurs, it gives us, of course, $1/(m+2)$.)

Another surprise is that the Rule reappears in Bayesian analysis. The assumption of independence and a flat distribution can be replaced by exchangeability (section 4.4) to gain a much more flexible version, which in turn can lead to a more subtle analysis of the applicability of the Rule. Again, Zabell 1989 is an excellent starting point.

5.1.2 The continuous case of the Principle of Indifference

The Principle of Indifference can be simply extended to the continuous case. Generally speaking, the probability that a given parameter takes a value in a given interval is equal to the size of that interval divided by the range of the parameter. For a more concrete example, consider (yet again) a wheel of fortune. Assume that we have no reason for believing that the pointer will stop at one point position rather than another (the pointer may halt at any point on the circumference, and we believe it to be uniformly lubricated, balanced, and so on). Consider the probability that the pointer will fall in an arc one quarter of the circumference of the wheel (say, in the third quarter, as in Figure 5.3). Assuming the circumference to have a unit length, the length of the arc is 1/4. The Principle of Indifference tells us that the probability of the arrow landing in this quarter (or any quarter) is 1/4.

Most generally, consider any continuous parameter A which ranges over an interval $[a, b]$, and for which in light of our present knowledge we have no reason to believe that it will take one value rather than another. Then, according to the Principle of Indifference, the probability that A lies in the sub-interval $[c, d]$, $p(c \leq A \leq d)$, is

$$\frac{|d-c|}{|b-a|}.$$

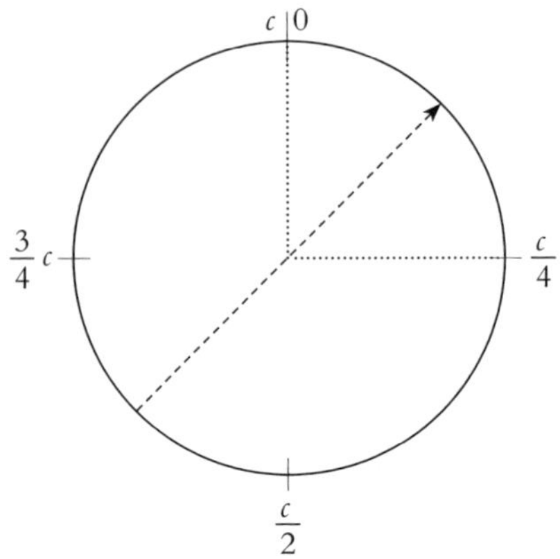

Figure 5.3 The rule of indifference on a wheel of fortune

5.2 Problems with the Principle of Indifference

The Principle of Indifference is not widely accepted, at least among philosophers. The first problem is with the Rule of Succession, which certainly appears to solve the problem of induction far too easily. There are also problems with the application of the Principle itself, since it seems to lead to contradictions.

5.2.1 Problems with the Rule of Succession

The most famous and perhaps impressive use of the Rule was Laplace's calculation of the probability that the sun will rise tomorrow, based on its having risen (on the authority of the Bible) for five thousand years. Things are never as clear as common lore would teach us, since what Laplace actually said was:

> Placing the most ancient epoch of history at five thousand years ago, or at 1826213 days, and the sun having risen constantly in the interval at each revolution of twenty-four hours, it is a bet of 1826214 to one that it will rise again to-morrow. But this number is incomparably greater for him who, recognizing in the totality of phenomena the principal regulator of days and seasons, sees that nothing at the present moment can arrest the course of it.
>
> (Laplace 1814: 19)

While a calculation of this sort does seem odd, it does not seem to be a reason to reject the Rule out of hand. There are other problems, however, that are more serious than just seeming odd.

Venn pointed out that while there are many cases where the answers given by the Rule seem reasonable, there are others where it does not:

Laplace has ascertained that, at the date of the writing of his work, one might have safely betted 1826214 to 1 in favour of the sun's rising again. Since then, however, time has justified us in laying longer odds. De Morgan says, that a man who, standing on the bank of a river, has seen ten ships pass by with flags, should judge it to be 11 to 1 that the next ship will also carry a flag. Let us add an example or two more of our own. I have observed it rain three days successively,—I have found on three separate occasions that to give my fowls strychnine has caused their death,—I have given a false alarm of fire on three different occasions and found the people come to help me each time. In each of these latter cases, then, I am to form an opinion of just the intensity of 4/5 in favour of a repetition of the phenomenon under similar circumstances. But no one, we may presume, will assert that in any one of these cases the opinion so formed would have been correct. In some of them our expectation would have been overrated, in some immensely underrated.

(Venn 1876, chapter VII, section 8, p. 180)

Venn addresses the obvious reply that his examples represent a misapplication of the Rule: he contends that this reply makes the application of the Rule subjective, given that it is dependent on the knowledge of the person applying it. He further points out that in this case the Rule gives no guarantee that it will give the right answer, and is therefore of dubious utility.

There is also the question of the appropriateness of the urn model needed to prove the Rule. First, we have to be sure we have identified the correct model, by providing the correct amount of colours of balls which correspond to the basic possibilities. Secondly, there is the question of the appropriateness of the drawing of balls from an urn with equiprobability: the assumption of equiprobability is very restrictive (for example, there will be probabilistic cases we cannot deal with—for a discussion see Zabell 1989). We return to this in 5.3.3.

These considerations give us good reason to doubt the soundness of the Rule. As we shall see in the following sections, the problems of the Rule mirror (and in fact are) the same problems that afflict the Principle of Indifference. We shall also see that these problems are very severe.

5.2.2 The paradoxes

The Principle of Indifference is famous for generating paradoxes. We will sum up the major classes of these paradoxes in the following sections.

5.2.2.1 The discrete case One of the simpler examples can be found in Keynes 1921 and is known as the Bookmark paradox; it raises problems for the discrete form of the Principle of Indifference. In the library at Prokop's university there are three colours of bookmarks (red, green, and blue, with exactly one bookmark per book). Suppose that, to relieve the tension of so much studying, he decides to go blindfolded into the library and pick out a book at random. Assuming that he is not arrested in the process, we have no more reason to believe that the bookmark will be red than not, assuming the alternatives of red and not-red. So, the Principle of Indifference implies that the probability of selecting a red bookmark is 1/2. But by similar reasoning the probability of getting a blue bookmark is also 1/2, as is the probability of getting a green one. But this violates the probability calculus: the probability of mutually exclusive and exhaustive alternatives must sum to 1.

5.2.2.2 The paradoxes—the continuous case The discrete case, as we shall see later, is fairly simple to deal with (although only on a very limited basis). Discrete probabilities, however, are rare in science, where we usually make use of continuous quantities. Temperature, mass, volume, length, and other fundamental quantities are measured using the continuum. But when the Principle of Indifference is applied in these cases inconsistencies arise that are much more difficult to deal with.

The wine-water paradox is one of the most famous examples of inconsistencies generated by the continuous form of the Principle. Expositions may be found in von Mises (1957: 77) and Gillies (2000: 37–49). Von Mises tells us he follows Poincaré in calling the paradox 'Bertrand's paradox', while Keynes (1921: 48–9) discusses a similar paradox which he attributes to von Kries.

Jarda goes to a pizzeria with Prokop ('It's a real pizzeria, the owners are émigrés from Prague!'). It's a hot summer evening, and Jarda wants a beer. But it's a real Czech pizzeria, and the beer is expensive and not very good. Luckily, they have some authentic lip-puckeringly sour red wine. Prokop proposes mixing wine with water to make a refreshing, yet alcoholic, beverage. Jarda reluctantly agrees. Prokop mixes the wine while Jarda orders a pizza with maize and sardines. Prokop then tells Jarda, truthfully, that the mixture contains no more than three times as much of one liquid as the other. Naturally, they jump at the chance to apply the Principle of Indifference, so they pull out their pens and start to write on the paper tablecloth so kindly provided by the restaurant. Prokop suggests that Jarda calculate the probability that the ratio of wine to water is less than or equal to 2. Jarda's knowledge does not extend beyond the inequality $1/3 \leq \text{wine/water} \leq 3$. He therefore has no reason for thinking that the ratio is one value rather than another. Hence applying the continuous form of the Principle of Indifference, he gets that $p(\text{wine/water} \leq 2) = \left|2 - \frac{1}{3}\right| / \left|3 - \frac{1}{3}\right| = \frac{5}{8}$. The waiter, looking over Jarda's shoulder, asks, 'What about the reciprocal ratio of water to wine? What's the probability the ratio of water to wine is greater than or equal to 1/2?' Jarda locates a fresh area of the tablecloth, and again applies the Principle of Indifference, getting $p(\text{water/wine} \geq 1/2) = \left(3 - \frac{1}{2}\right) / \left(3 - \frac{1}{3}\right) = \frac{15}{16}$. The waiter says 'Hmm' and returns to the kitchen.

These two results contradict one another, since wine/water ≥ 1/2 is the same state of affairs as water/wine ≤ 2. The Principle of Indifference has given different probabilities for logically equivalent propositions in violation of the probability calculus. (And so has generated a contradiction. A paradox is more or less an argument from plausible premises to a conclusion we don't like. Jarda doesn't like this conclusion, so he decides it's a paradox, instead of a *reductio ad absurdum*.)

5.2.2.3 The paradoxes of geometric probability (Bertrand's paradox) Another set of examples of inconsistencies generated by the Principle of Indifference are the 'paradoxes of geometric probability'. Keynes (1921: 51) attributes the following example of such a paradox to Bertrand, and Neyman (1952: 15) refers to 'the so-called Bertrand's problem'. I follow Borel's exposition (1950: 87). The problem is to determine the probability that a chord drawn at random onto a circle of radius *r* will be shorter than the side of an equilateral triangle inscribed in the circle.

There are at least three ways of calculating the probability using the Principle of Indifference, illustrated in Figures 5.4, 5.5, and 5.6. In the first, we might label the point at which one end of the chord falls *A*, and apply the Principle of Indifference to the position where the other point falls (F or G in Figure 5.4). Consider the line DE tangent to the circle at the fixed end point *A*. This and the chord will form an angle ∠FAD (or ∠GAD). If this angle is between 60 and 120 degrees, the chord AG will be longer than the side of an inscribed equilateral triangle ABC, otherwise it will be shorter (AF). So, applying the Principle of Indifference, the chance that the chord will be shorter than the side of an inscribed triangle is 60/180 = 2/3. (Indifference is being applied with respect to the chord's angle of intersection at the circumference: the probability is, as can be seen from Figure 5.4, obviously 2/3.)

In the second case (Figure 5.5), the Principle of Indifference can be applied to where the midpoint of the chord falls with respect to a fixed

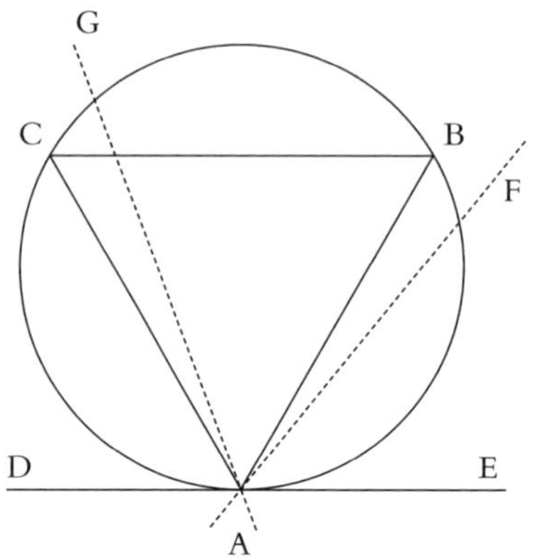

Figure 5.4 Bertrand's paradox, first case

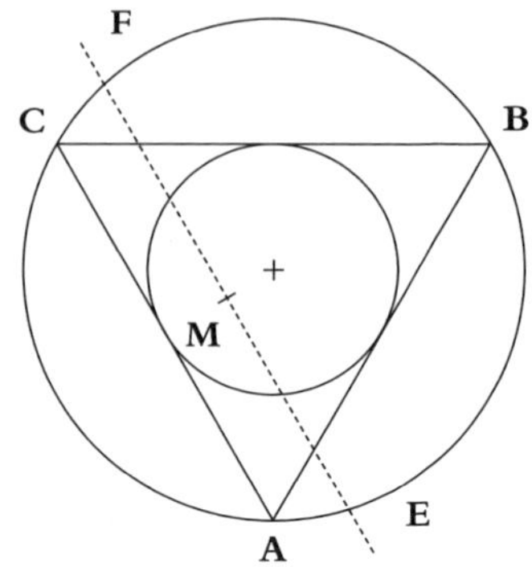

Figure 5.5 Bertrand's paradox, second case

direction. The chord must fall in *some* direction. Consider an inscribed triangle ABC with side BC parallel to this direction. We can determine the length of the chord as follows: a line drawn perpendicular to BC from the centre of the circle to the circumference of the circle will have its midpoint where it intersects the triangle (P). If the midpoint of the chord (M, for the chord FG) falls between the midpoint of the perpendicular line and the circle, the length of the cord will be less than that of the side of the triangle (as is FG). Similarly, if the midpoint falls less than half the distance, it will be longer (as is DE). So the probability of the chord's length being less than the side of an inscribed equilateral triangle is 1/2. (Indifference is applied to the radius: obviously the answer will be 1/2.)

A third way of calculating the probability of the length of the chord being shorter than the side of an inscribed equilateral triangle is shown in Figure 5.6. Consider a circle inscribed in the inscribed triangle, and denote the midpoint of the chord as M. Since the inscribed circle has radius $(1/2)r$, if the midpoint falls inside that circle, the chord (FE in the figure) will be longer than the side of an inscribed triangle. The area of the smaller circle G is 1/4 of the area of the larger circle, and so the probability that the chord will be smaller is 3/4. (Indifference is applied with respect to total area of the circle.)

Hence the Principle of Indifference assigns at least three different probabilities to one proposition, and, once again, generates a contradiction.

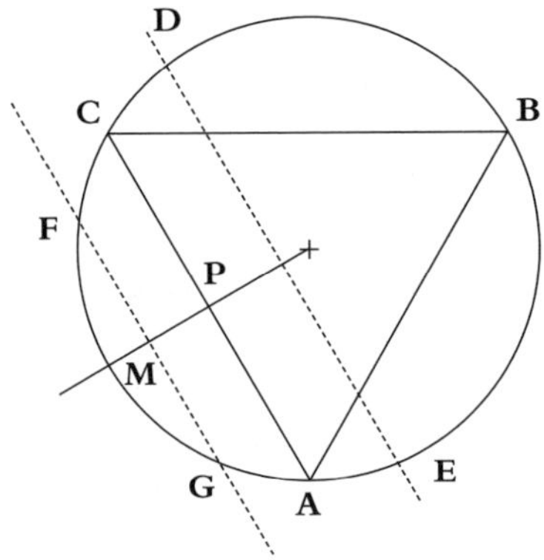

Figure 5.6 Bertrand's paradox, third case

(I have given the traditional explication of the paradox. Marinoff 1994 carefully shows that things are much more complicated than they seem. This is a point we return to in 6.4.2 and 6.4.3.)

5.2.2.4 Linear transformations and the Principle of Indifference Bertrand's paradox and the wine/water paradox show that the Principle of Indifference is not invariant with respect to non-linear transformations such as inversion. As Keynes (1921: 51) pointed out, these paradoxes arise since, in general, $|d-c|/|b-a|$ is not the same as $|f(d)-f(c)|/|f(b)-f(a)|$.

The classical theory seems to face insurmountable difficulties. This did not, of course, stop philosophers from trying to surmount them. We will address the most well-known attempted solutions associated with Keynes and Carnap, since it serves to introduce the modernized version of the classical theory along with its associated triumphs and failures.

5.3 Keynes's Logical Interpretation

Although the two are usually kept separate, I will take the logical interpretation of probability to be an updated version of the classical theory: we can take the logical interpretation to be the classical interpretation subjected to a linguistic turn. Thus, I'm going to take the classical/logical view as assigning equal probabilities to propositions (absent any information leading us to assign them otherwise). The propositions, of course, represent the basic possibilities.

The first fairly worked out logical interpretation of probability was provided by Bernard Bolzano in the first half of the eighteenth century. In fact, more or less the same form of logical probability was proposed by almost every Central European philosopher of note from Bolzano to Wittgenstein (for some details see Childers and Majer 1998).

One way of putting the logical interpretation is that it takes the notion of probability to be that of partial entailment. The basic idea is that the number of propositions entailing a statement, when put in a ratio to total propositions, gives the probability of that statement. Each proposition counts as one, and so each is assigned probability $1/n$, where n is the total number of propositions.

The Principle of Indifference (somewhat) logically formulated states that in situations where we have n mutually exclusive and exhaustive hypotheses $h_1, \ldots, h_n$, and no evidence favouring any particular hypothesis, we should assign equal probability to each of them, that is, for each hypothesis h_i, $p(h_i) = 1/n$. Obviously, since it's a ratio of the appropriate type, logical probability satisfies the Kolmogorov axioms. Partial entailment is then defined in terms of conditional probabilities—but we shall leave the details to section 5.4.1. We will explore the logical interpretation first in the work of John Maynard Keynes, and then in that of Rudolf Carnap.

5.3.1 The discrete case and the justification of the Principle of Indifference

The first notable attempt to deal with the paradoxes by employing the resources of the new mathematical logic of Russell and Whitehead was made by John Maynard Keynes in his 1921 *A Treatise on Probability*. Keynes was a student of Russell and Moore, and a friend of Ramsey and Wittgenstein, and so well placed to attempt this project.

The paradoxes arise from changes of description. So, one way to avoid the paradoxes is to make sure that the descriptions can't change: the obvious way to do this is to apply the Principle of Indifference only to indecomposable alternatives. Keynes took this route, restricting the application of the Principle of Indifference to those cases in which the hypotheses we are indifferent between are not further decomposable, or, in Keynes's terminology, 'divisible'. Keynes attempted to specify what decompositions he had in mind: a sentence is decomposable if it is equivalent to a sentence which is a disjunction of statements which are (1) exhaustive, (2) mutually exclusive, and (3) have positive probability (1921: 65). The problem with this account, as Keynes acknowledged, is that any sentence can always be 'decomposed' in the wrong way: statements of the form A are equivalent to $((A \wedge B) \vee (A \wedge \neg B))$, where B is any statement. This satisfies conditions (1)–(3), and so it would seem that Keynes's solution will not work. However, Keynes points out that the sentences are not of the same form. Keynes does not tell us what sameness of form is: working this out will have to wait for Carnap, and section 5.4.1.

But even if we can work out a reasonable notion of decomposability, serious difficulties will remain. Even assuming that there *are* basic possibilities of the kind necessary is an exceptionally strong philosophical assumption, seemingly equivalent to logical atomism. Further, if we have not actually discovered the true possibilities over which to distribute probability, we will always get the wrong answer. Another way of putting the difficulty is that logical probability is *language-dependent*: if you change the language, you get different answers. But linguistic divisions rather obviously need not reflect real divisions. Therefore, the logical interpretation may tell us something about how language behaves, but not necessarily anything about how anything non-linguistic behaves.

Keynes had, from a contemporary point of view, a somewhat eccentric account of how to ensure that we did get the proper basic possibilities. He employed Russell's distinction between knowledge by acquaintance and knowledge by description to define 'direct knowledge'—propositional knowledge based on being directly acquainted with something. He further took this knowledge to be indubitable. We could also perceive, according to Keynes, second-order ('secondary') relations between propositions given by direct knowledge, including probabilistic relations. Keynes also employed G.E. Moore's intuitionist account of ethical knowledge (which is obviously related to Russell's account of direct acquaintance) to argue that probabilities could not be defined in terms of any simpler notions. With these notions, he then argued that in certain cases we could make direct judgements that the Principle of Indifference was applicable (in particular, we could judge that there was no other proposition that affected their probability—thus we would be indifferent to the propositions).[1] If what has been said so far is correct, then we have an independent justification for assigning equal probabilities: directly perceiving the probability relations

[1] An interesting result of this account is that in some cases it is not possible to order all propositions by their probabilities: in some cases we will not have direct knowledge of the probability relations between two propositions, and so can only give a weaker characterization of the relation of the propositions to some, but not all, other propositions.

between propositions, being based on knowledge given by direct acquaintance, is not language-dependent, and is, moreover, indubitable.

However, there is very good reason to think that Keynes was wrong. Moore's intuitionism is very much out of favour, since there are strong arguments against it. Keynes's intuitionism is subject to even stronger objections, and so his programme is generally (actually, almost universally) taken as a failure. The failure makes for an interesting philosophical story, but I will leave the telling to another occasion.

5.3.2 Keynes on the continuous case

Another serious problem is that it is not clear that Keynes has a solution to the paradoxes in the general case. For the wine/water paradox he argues that the Principle is simply inapplicable (1921, chapter 4, section 23), 'because there is no range which does not contain within itself two similar ranges'. No matter how finely the possibility space is partitioned, the remaining alternatives can always be further decomposed. Hence, the Principle of Indifference is not applicable in these cases (Keynes 1921: 67).

But Keynes seems to have believed that some of the paradoxes of the continuous case could be overcome with his modified Principle of Indifference. He apparently thought that the paradoxes in the continuous case could be avoided if the parameter of interest were restricted to a finite number of values in m intervals on the real line, and if m were then allowed to tend to larger values:

> Suppose, for instance, that a point lies on a line of length $m.l.$, we may write the alternative 'the interval of length l on which the point lies is the xth interval of that length as we move along the line from left to right' $\equiv \mathrm{f}(x)$; and the principle of indifference can then be safely applied to the m alternatives $\mathrm{f}(1), \mathrm{f}(2), \ldots, \mathrm{f}(m)$, the number m increasing as the length l of the intervals is diminished. There is no reason why l should not be of any definite length however small.
>
> (Keynes 1921: 67–8)

It is unclear how this would help. We must first determine which value of m we take to be indivisible: surely any such choice will be arbitrary; indeed, it would seem to be just wrong. Second, it seems that if m does tend to infinity, we once again are dealing with a continuously valued parameter, and so the paradoxes will re-emerge (since the probabilities defined over the parameter will not be invariant under transformation—an example sufficient to demonstrate this can be found in Howson and Urbach 1993: 60). Also, it is difficult to interpret Keynes on this matter, because he then in the next paragraph seems to offer a different solution, based on how the cases in the geometric paradoxes are characterized (taking the Principle of Indifference to apply not to the chord, but to the shape used to determine a solution, perhaps in the spirit of the solution we will encounter in 6.2.2). However fascinating Keynes exegesis may be, we can conclude that he did not offer a workable solution to the paradoxes in the continuous case. But many sciences work almost exclusively with the continuous case. Hence, unless a solution is found, the logical interpretation will be of very limited value for science.

5.3.3 Keynes on the Rule of Succession

Keynes strongly criticized the Rule of Succession on a number of grounds. Zabell 1989 replies to some of his criticisms, but one of Keynes's criticisms deserves particular attention: the suitability of the urn model as one of inductive inference. Keynes points out that the urn model is a very restrictive assumption, and in general will not hold. For suppose we wish to estimate the relative frequency of the occurrence of an event in an unlimited series of throws. Then the Rule of Succession can be derived if we assign all possible proportions of m draws from the urn of, say, the white ball, over n total trials. Keynes argues that there is no reason to believe that these assumptions hold in any particular situation in which we are conducting observations. And obviously, once again, we would have to assume that we have got the actual constituents of the urn correct—that is, we have to assume that we have correctly modelled our phenomenon. In this case, Keynes's means of justification will probably not do, as we could not make the direct judgements necessary.

Keynes's book was a major advance in the study of the foundations of probability. It contains fascinating and deep discussions of many foundational issues that are still relevant. But those are not central to our exposition of the logical interpretation, and so we now turn to the next, and last, great attempt to found the logical interpretation, undertaken by Rudolf Carnap.

5.4 Carnap

Carnap was, of course, a member of the Vienna Circle, which rejected metaphysics as meaningless. Members of the Circle agreed with Hume that all meaningful statements are either logical or empirical, corresponding, respectively, to the a priori and the a posteriori. (They also rejected Kant's claim as to the existence of synthetic a priori truths, hence equating the a posteriori with the synthetic, and the a priori with the analytic.) Accordingly, they stressed analyticity as the basis of all a priori knowledge (a logical statement is said to be analytic if it is true solely in virtue of its meaning). For Carnap, the purpose of philosophy was to divide the true statements of science into these categories ('*Philosophy is the logic of science*, i.e. the logical analysis of the propositions, proofs, theories of science' (Carnap 1934: 54–5); 'But what, then, is left over for *philosophy*, if all statements whatever that assert something are of an empirical nature and belong to factual science? What remains is not statements, nor a theory, nor a system, but only a *method*: the method of logical analysis' (Carnap 1932: 77)).[2] As a part of this programme of analysing science, Carnap attempted to provide a foundation for an inductive probabilistic logic by showing that the axioms of inductive probability are analytic: 'all principles and theorems of inductive logic are analytic' (Carnap 1950: v). This last statement may be confusing: for there is no doubt that the *theorems* of the probability calculus are analytic (that is, if anything is analytic, mathematical theorems are). There is, however, a doubt as to whether substantial inductive *principles* are analytic—that is, whether we can derive, a priori, a solution to the problem of induction. As we shall see, Carnap failed to find such a solution. In fact, since he so convincingly failed to do so, we can draw an instructive lesson.

5.4.1 The logical foundations of probability

According to Carnap, deductive relations are analytic in the sense that whatever is contained in the conclusion of a valid deductive argument is also contained in the premises. Carnap wanted a similar notion of containment for probabilistic inductive logic. In Figure 5.7, on the left-hand side, h entails e because h contains e (say, all models that make e true also make h true). On the right-hand side, h and e only share some models, and so h only partially contains, and hence only partially entails, e.

Carnap developed logical machinery necessary for a treatment both of deductive logic and what he took to be inductive logic. We shall briefly survey this. The simplest sentences are those of a primitive language with just one predicate, call it A, n distinct individuals, $a_1, \ldots a_n$, and the sufficient set of connectives $\neg$ and $\wedge$ (the others being defined as usual). We can

[2] Of course, the story of the Vienna Circle is vastly more complicated than I'm letting on. Still, it will do.

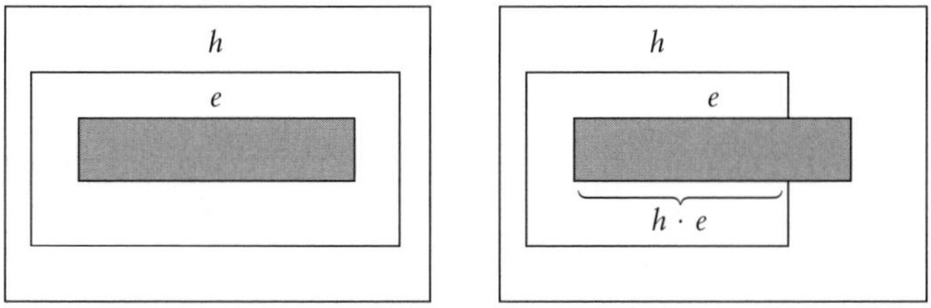

Figure 5.7 Full and partial entailment

represent all possible states of affairs describable in this language by means of the 2^n conjunctive sentences $\pm A(a_1) \wedge \pm A(a_2) \wedge \ldots \wedge \pm A(a_n)$ (where $+A(a_i)$ is $A(a_i)$ and $-A(a_i)$ is $\neg A(a_i)$). Carnap called these sentences *state-descriptions*. A state-description can be thought of as a possible world: the status of each individual in that world is described. Each state-description is incompatible with all the other state-descriptions, and so they completely divide up the possibilities (or at least, the possibilities relative to the descriptive strength of a language). They can thus serve as Keynes's indecomposable propositions. An example: you have a world with two books, and two colours, red and not-red. There are four state-descriptions generated by the language describing this world: both books are not-red, both books are red, the first book is red, the other not, the first book is not-red, the other red. Obviously this language, and any language with only one predicate, is very limited: it's used for purposes of illustration only.

Carnap defined the *range* of a statement as the set of state-descriptions that are compatible with it. Deductive logic can be described in terms of ranges. A sentence is a tautology if and only if it holds in all possible worlds (that is, it is compatible with the truth of every state-description). It is a contradiction if and only if it does not hold under any state-description. A statement h implies e if and only if the range of e is included in h (that is all state-descriptions compatible with h are also compatible with e). Finally, two statements are equivalent if and only if they have the same range.

Carnap interpreted inductive logic as a probability measure of ranges. This can be seen by defining a function m such that

1. $\Sigma_i m(\mathrm{P}_i) = 1$, where the P_i are the state-descriptions.
2. If h is not logically false, $m(h)$ is equal to $\Sigma_j m(\mathrm{P}_j)$, where the P_j are the state-descriptions that make h true (i.e. $\{\mathrm{P}_j\}_j$ is the range of h).
3. If h is logically false, $m(h) = 0$.

and a function c such that

4. $c(h \mid e) = m(e \wedge h)/m(e)$.

Clearly, m is a probability, and c is a conditional probability. $c(h \mid e)$ is a measure of the inclusion of the range of e in the range of h, and so is a measure of partial entailment. In virtue of the definition of the measure, any theorem which follows from that definition will be analytic (in Carnap's view). This might seem to justify Carnap's claim that inductive logic is analytic.

However, the statement that $c(h \mid e) = r$ for a given h and e and a particular r is in general *not* analytic, since this is dependent on the values the function m assigns, and (almost) any value between 0 and 1 will satisfy the definitions. So the axioms alone cannot ground a logical interpretation of probability, since they do not uniquely determine probability assignments, and in particular not symmetric ones. This can be easily seen by taking a language with one individual and one predicate: you can assign any number between 0 and 1 to $p(+Aa)$ as long as when added to $p(-Aa)$ it sums to 1. (A similar argument can be made for conditional probabilities.)

Hence Carnap sought additional restrictions on probability measures in order to restrict admissible probabilities. Rules governing the assignment of particular probability values within the earlier constraints of 1, 2, and 3 are called by Carnap inductive methods. However, his investigations led him to the discovery of infinitely many inductive methods, the λ-continuum. From this continuum of inductive methods, he concentrated on two measures, $m^\star$ and $m^\dagger$.

5.4.2 *The continuum of inductive methods*

In his *Logical Foundations* Carnap considered only two conditional probability measures, $c^\dagger$ and $c^\star$ (their associated unconditional probability measures are $m^\dagger$ and $m^\star$). Both are obtained by imposing symmetry conditions: the first by assigning state-descriptions equal probability (considered earlier), and the second by assigning structure-descriptions equal probability. (Structure-descriptions are described later.) $c^\dagger$ is the measure favoured by Łukasiewicz, Wittgenstein, and Keynes. $c^\star$ is the classical measure favoured by Laplace.

$c^\dagger$, as I just said, is the conditional probability obtained earlier by applying the Principle of Indifference to state-descriptions. Since in our simple language there are 2^n state-descriptions, if we assign equal probability to each of the state-descriptions, they would each have probability $(1/2)^n$. This measure turns out to be insensitive both to sample size and to information about the proportion of a_i's possessing A. The probability of an event turning out in any particular way given that other events have turned out in some way is always 1/2, no matter how many events have been observed. Such a measure cannot be inductive, since it does not provide for learning from experience, and for this reason Carnap rejected it (1952: 38). (Wittgenstein, who in the *Tractatus* takes $c^\dagger$ as his measure, was well aware of this result: but he was not aiming to construct an inductive logic—see Childers and Majer 1998.)

The measure $m^\star$ is based on equiprobability of structure-descriptions. A structure-description is a description of all the ways the property can be distributed among the individuals: this is done by forming a disjunction from all state-descriptions with the same number of predicates. For example, in the simple case of a language with one predicate and two individuals, there are three structure-descriptions:

$$A(a_1) \wedge A(a_2), (A(a_1) \wedge \neg A(a_2)) \vee (\neg A(a_1) \wedge A(a_2)), (\neg A(a_1) \wedge \neg A(a_2))$$

Assigning equal probability to all the structure-descriptions is equivalent to the Laplacian Principle of Indifference, and so Carnap was able to derive the Rule of Succession. Since this measure is sensitive to the size of the sample, it seems more satisfactory than one based on the equiprobability of state-descriptions.

Carnap later (in his 1952 *The Continuum of Inductive Methods*) developed a general means for classifying measures sensitive to sample and population size using a positive real-valued parameter λ (and in which $m^\dagger$ and $m^\star$ are at opposite ends of the spectrum of values of $c^\dagger$). The λ-continuum of inductive methods is that set of methods determined as follows: Given that m individuals in a sample of size n have A (denote this as $rf(m, n)$), the probability of the next observation being an A, A_{n+1} is

$$c_\lambda(A_{n+1} | rf(m,n)) = \frac{m + \frac{\lambda}{k}}{n + \lambda},$$

where k is the number of attributes (predicates). This classification gives a range of proposed measures: for example, if $\lambda = 0$, the probability of the next event having A is simply the observed relative frequency of A's (the so-called 'straight rule'). If $\lambda = 0$ we get $c^\star$. In a binomial language, $k = 2$; $\lambda = 2$ gives the Rule of Succession.

Carnap seems to have hoped to justify setting $\lambda = 2$. He failed, however, in providing an independent justification for doing so. Since there is no reason to pick one value of λ over another, there seems little motivation to stick with the limitations of the Carnapian framework. These limitations seem rather severe. For example, it's not even clear if all inductive methods can be represented in Carnap's framework (counter-inductive methods cannot). And worse, there is no straightforward generalization to the continuous case, making Carnapian logical probabilities unsuitable for science. Carnap's programme is widely regarded as coming to an end in the early 1970s, and with it the notion of a logical interpretation of the probability calculus.

5.5 Conclusion

Logical probability has enjoyed a revival as of late. Two different strands of work are responsible for this. The first comes from work in mathematical logic by Paris and Vencovská and their collaborators. They have made a remarkable leap forward by applying the notion of automorphisms to make precise the symmetries required for logical probabilities. An automorphism, as the name suggests, is a mapping of an object onto itself that preserves structure. A standard example is the rotation of a triangle by 360 degrees: the triangle remains unchanged. Mathematicians use groups of automorphisms to describe symmetries (the classic account is Weyl 1952; a more recent philosophical contribution is Guay and Hepburn 2009). It seems therefore natural to apply this tool to logical structures. The resulting account of probabilities is quite sophisticated (and rather technically demanding). A good place to start is Paris and Vencovská 2011 and 2012. The second strand in the revival of logical probability comes from physics, and is the subject of the next chapter.

6

The Maximum Entropy Principle

The Maximum Entropy Principle can be seen as a natural continuation of the Carnapian programme in Chapter 5. This Principle came to prominence in the work of E.T. Jaynes. Jaynes updates the Principle of Indifference by building on Claude Shannon's astonishing 1948 paper.[1] Shannon developed a quantitative measure of information which can be used to determine symmetrical probability distributions. Proponents of the Principle hold that it avoids many, if not all, of the difficulties of the logical interpretation. However, we shall see that despite its many attractive features, it suffers from the same flaws, albeit in a more sophisticated form.

6.1 Bits and Information

Prokop is contemplating adding more memory to his computer so that his favourite graphics programme will run smoothly. Memory is sold in units of gigabytes at the time of writing. Bytes are made of bits, eight of them. A bit is a single unit of memory, that is, a space for storing a 1 or a 0. A gigabyte is a lot of bytes.

We can think of a bit as a unit of information. While it might be tempting for some to think of a bit as, say, a cell (a capacitor and a transistor) on a RAM chip having a charge or not, we will concentrate on what the cell having a charge represents—what it means. A bit can represent a switch being on or off, or a coin coming up heads or tails, or any binary state of affairs. Abstractly, we can think of a bit as an indicator variable. Despite

[1] Why 'astonishing'? Because Shannon produced, in nearly complete form, an entire field.

the somewhat misleading name, an indicator variable is a function that takes 1 or 0 given the occurrence of its argument. For example, it might take 1 if the toss of a coin comes up heads, 0 otherwise.

An indicator variable tells us that one particular thing is in one state or not. Consider the amount of information a single such variable *A* can give: it says that the state is either 1 or 0, on or off, heads or tails. If we have two such variables *A* and *B*, we have double the amount of potential information, since there are now four possible states instead of two (*A* is 1, *B* is 1; *A* is 1, *B* is 0; *A* is 0, *B* is 1; *A* is 0, *B* is 0: the first switch is on, so is the second, etc., the first coin came up heads, so did the second, etc.). *C* quadruples the amount of information (to see this, think of the number of rows in the truth table for *A*, *B*, *C*) and so on.

Each time we add an indicator variable we exponentially increase expressive power, for we can describe exponentially more possibilities. Instead of, for example, describing a single coin toss, we can now describe three, and instead of only two possible outcomes, we now have eight. Describing the outcome of a single coin toss requires one bit, describing three coin tosses requires three, and so on, as our example makes clear.

More abstractly: with one indicator variable *A* we can describe two states, with two, *A* and *B*, we can describe four states, with three, *A*, *B*, and *C* we can describe eight: where *n* is the amount of variables, the number of states is 2^n. Think of the possible state space as the space of possible configurations. Every time we add a variable, it grows much larger. Mathematically, we say that information varies exponentially with a linear increase in indicator variables. The natural measure to use in this case is the logarithmic measure, which will then make information linear as well (see Appendix A.0.5 if logarithms make you nervous or if you need a quick refresher). So, for one binomial random variable *A* with two potential states, the amount of information is $\log_2 2 = 1$; for *A*, *B* it is $\log_2 4 = 2$; for *A*, *B*, *C* it is $\log_2 8 = 3$; and so on. We now have a reasonable-seeming measure of information, $\log_2 n$, where *n* is the number of states. (A more complicated version would allow the indicator variable to indicate more than two states. This can be done by changing the base of the logarithm.)

So far, so good, but this is only because we have been dealing with information defined in bits as used by computer manufacturers. We can aim for a more general notion of information, a notion of potential information that arises from the chance that an indicator variable will take a certain value. This notion would measure information's relation to certainty: learning something you are certain of gives you no new information, but learning something you thought was false gives you a lot of information. Using probability to model this: the greater the probability of an event, the less potential information.

One path to a measure of information, or better, informativeness, is to consider sending a coded message about the outcome of an experiment. As an experiment we will use the outcome of a role of an eight-sided die (we'll see shortly why we use eight as opposed to six). The message will be sent in a string of 0's and 1's, i.e. bits. There are numerous ways to encode the outcomes, of course; the following table gives an example:

1	0
2	10
3	110
4	1110
5	11110
6	111110
7	1111110
8	11111110

Suppose we don't know the outcome, but think each one equally likely. Then the expected length of the message for this coding scheme is

$$1/8 + (1/8)2 + \ldots + (1/8)8 = 3^{1}/_{2}.$$

A more compact coding scheme would be:

1	111
2	110
3	101
4	100
5	011
6	010
7	001
8	000

Here the expected length is, obviously, 3. The relation with probability is as follows: to record eight equally likely outcomes we need three bits. This can be determined by taking the logarithm of the inverse of the probability (the needed number of equally likely outcomes). In this case, we need the number to which we must raise 2 to get eight outcomes, that is, is 3. The general formula

$$\log_2 \frac{1}{p(x_i)} = \log_2 p(x_i)^{-1} = -\log_2 p(x_i).$$

serves as an optimal estimate of the average length of the optimally compact code, where x_i encodes an outcome, i.e. is a random variable.[2] Following common practice, we shall now denote this measure as $I(x_i)$.

As can easily be seen, this definition of information reduces to our preceding definition when the random variables are independent and have equal probability. But we have more than a measure of bits, as can be seen from considering the graph of $I(x_i)$ (Figure 6.1). If the outcome is certain, then the information carried by the message is 0: you already knew the

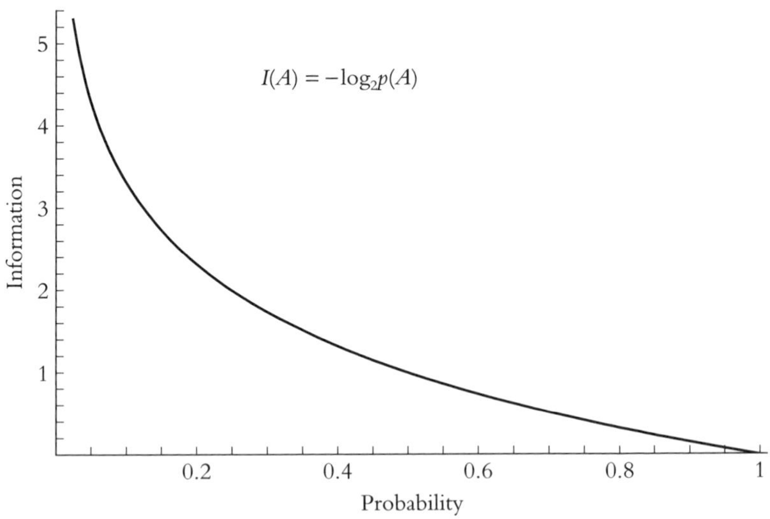

Figure 6.1 Probability and information

[2] We used the eight-sided die instead of the usual six-sided one to avoid the following complication: $\log_2 6$ is an irrational number. This can be easily—but will not be—proved. So the $\log_2 p(x_i)$ measure is an ideal.

event would happen, and so it's not informative at all. As the probability of the outcome approaches 0 the measure goes to infinity: the message gives much more information if the event that occurs is very unlikely. The 'middle', the mean, of the function is 1, the case where you just don't know one way or the other, i.e. $p(x_i) = .5$

The graph in Figure 6.1 should also make clear why $I(x_i)$ is also known as the surprisal value of A: if $p(x_i) = 1$, then there's no surprise, and so $I(x_i) = 0$. ('The sun will rise tomorrow!' It does. You say 'meh'.) But as $p(x_i)$ approaches 0, the surprisal value goes to infinity (shrieks of 'Oh my God!' by atheists, etc.). In the middle we have 'huh', neither expected nor unexpected.

6.2 The Principle of Maximum Entropy

$I(x_i)$ is itself a random variable, and so we can check the probability that it will give a certain value. That is, we can find its expectation (its mean—see Appendix A.3.2), denoted in the literature as $H(x_i)$. In the most general form, where the random variable can take n possible outcomes, and not just two,

$$H(A) = E(I(A)) = -\sum_n p(A_n)\log_2 p(A_n),$$

where A_n indicates that A takes the nth possible outcome.

Fascinatingly enough, the formula we have obtained as the expectation of information is the same for entropy found in physics, and hence $H(A)$ is referred to as the (Shannon) *entropy* of A. We will not discuss the relation between this quantity in our information-theoretic setting and in physics, but refer the interested reader to the references in Uffink 1996. However, part of the work Jaynes is famous for is his reformulation of statistical mechanics as dealing with uncertainty, and hence a reformulation of statistical mechanics not so much as a physical theory but 'as a theory of statistical inference, i.e. as a branch of logic or epistemology' (Uffink 1996: 224).

Determining the entropy of our ordinary binomial random variable is simple, since there are only two possible outcomes, $A = 1$ or $A = 0$. So the expectation in this case is

$$-p(A)\log_2 p(A) - [(1 - p(A))\log_2(1 - p(A))].$$

We can graph this function for values of p, as shown in Figure 6.2:

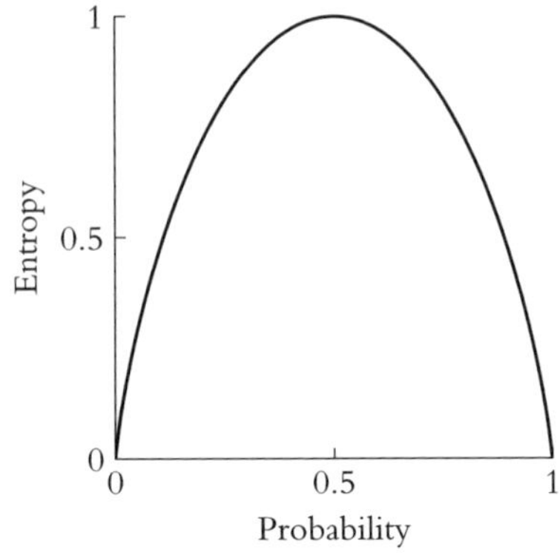

Figure 6.2 Probability and entropy

The peak of the function—the maximum of entropy—is when the probabilities are equal. This is usually interpreted as showing that the mean of the function, in this case .5, is where we have no information in favour or against the occurrence of a particular outcome of A. (Figure 6.2 can be found in Shannon 1948.)

It can be shown that entropy is at a maximum when the probabilities of the different possible outcomes are equal. Associating the maximum of entropy with the minimum of information, we get an updated version of the Principle of Indifference:

> *The Principle of Maximum Entropy*
> Assign probabilities in accordance with background knowledge. When you have no other information about the possible values that a set of random variables might take, assign probabilities that maximize the entropy of those variables.

As Jaynes, the first proponent of this Principle, put it: 'in making inferences on the basis of partial information we must use that probability distribution which has maximum entropy subject to whatever is known' (Jaynes 1957a: 623).

To explain the sorts of constraints background knowledge might place on the application of the Principle, suppose that you are tossing a die, and that it is ruled out, never mind how, that the die can come up 1 or 6. Suppose further that you have no other information about the die. Then, according to the Principle, you should assign the remaining possible outcomes probabilities that maximize entropy. In this case, that will be the equiprobable distribution, that is, each of 2, 3, 4, and 5 have $1/4 = .25$ probability.

Since this Principle requires that everyone with the same information adopt the same probability distribution, the interpretation of probability based on the Principle is sometimes called *Objective* Bayesianism. We shall avoid using this name, for reasons that will become clear in 6.4.

Before we turn to a discussion of how this theory fares with the problems faced by the Principle of Indifference, we will examine the continuous version of the Principle of Maximum Entropy. This has been of great interest to probabilists since it seems to offer hope in solving some of the most intractable problems of the logical interpretation of probability.

6.2.1 The continuous version of the Principle of Maximum Entropy

(This section is advanced, and requires some familiarity with the differential calculus. It may be skipped or skimmed, at least until you decide to devote yourself to the philosophical problems discussed herein.)

Our first task is to settle the form of the continuous Principle of Maximum Entropy. It might seem that we could just take the discrete version, and, following usual mathematical practice, apply it to larger and larger sets of alternatives (more and more fine-grained possibilities). As we let this process go to infinity, it seems that we would get (by analogy with the continuous form of the Principle of Indifference) the continuous form known as differential entropy:

$$-\int p(x)\log p(x)\,dx$$

But we do not: instead, we get an infinite sum. (A proof can be found in Cover and Thomas 2006: 247–8, who do not mention that the second term of their equation 8.29 goes to infinity as $\Delta \rightarrow \infty$.)

So we still need a continuous form. One (admittedly odd) strategy would be to simply ignore this inopportune result, and assert that differential entropy *is* the continuous form, because it looks like the discrete form. There are two reasons that most authors do not take this tack: first, differential entropy can be negative (Ash 1965: 237); second it is not invariant under change of variables (just as probability distribution functions generated by the continuous Principle of Indifference are not).

There are other quantities that are held to be the continuous form of entropy, the most popular of which is *relative entropy* (also known as Kullback-Leibler divergence):

$$\int p(x) \log \frac{p(x)}{q(x)} dx,$$

where $p(x)$ and $q(x)$ are probability density functions (see Appendix A.3.3). $p(x)$ is the distribution to be determined by the application of the Maximum Entropy Principle, and $q(x)$ 'is an "invariant measure" function, proportional to the limiting density of discrete points' (Jaynes 1968: 15). In other words, $q(x)$ is a flat distribution. One way of looking at relative entropy is as measuring the divergence between two probability distributions: if they are completely different, the divergence will tend to infinity, if the distributions are the same, it tends to 0.

Relative entropy has a number of 'nice' properties. In some cases it reduces to Shannon entropy, but is non-negative, and remains invariant under transformation. This means that the Principle of Maximum Entropy appears to both provide a flat distribution like the Principle of Indifference, as well as overcome the paradoxes associated with it. The next sections will explore whether it really does.

6.2.2 Maximum entropy and the paradoxes of geometric probability

Prokop and Jarda are in the garage of the apartment complex. There is a chalk circle on the floor in which are inscribed two triangles (Figure 6.3). Prokop is throwing wooden dowels at the circle, and calling out 'Longer!' and 'Shorter!'; Jarda is tabulating the results. The Reverend has looked in, cried 'Get thee behind me Satan!' and fled. ('Huh, it does sort of look like a pentagram'. 'Yeah, except it's got six sides'.) After Prokop's demonstration of the geometric paradoxes recounted in 5.2.3, namely, the multitude of possible answers to the question of the probability of a randomly drawn chord being longer than the side of a triangle inscribed in a circle, Jarda asked the obvious question: 'OK, so which answer is right?' This rather stumped Prokop, since he had only wanted to show that the Principle of Indifference led to contradictions. But Jarda is a practically-minded soul, and suggested setting up an experiment to settle the question. They set up a circle, and inscribed a couple of triangles to make data easier to eyeball, and set about testing the question. The data are in, and it's clear: the answer is that the chance of a chord falling within the circle having a length shorter than that of an inscribed triangle (or of a Star of David, which is much more convenient for measurements) is 1/2. The second explanation, then, is right, asserts Jarda. Prokop chews his newly grown van Dyke.

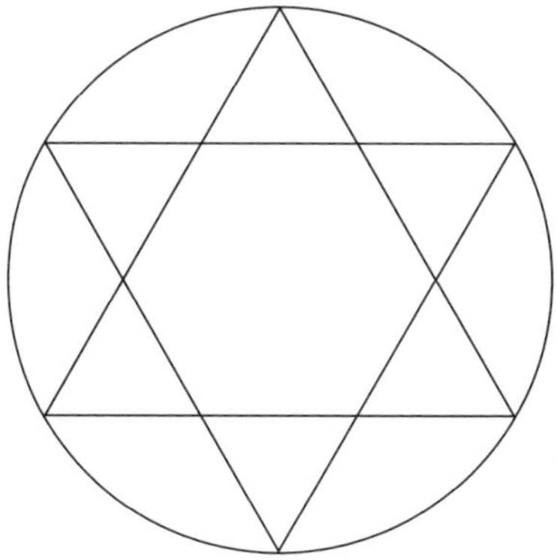

Figure 6.3 Jarda and Prokop's recreation of Jayne's experiment

Jaynes undertook this same experiment in 1973, and got the same result. Further, Jaynes claimed to have not only demonstrated which explanation is right, but also to have an explanation in line with the Maximum Entropy Principle, thus dissolving Bertrand's paradox. Jaynes helpfully rephrases the central problem of the paradox as one of choosing a parameter to which we should be indifferent. Is it 'the angles of intersections of the chord on the circumference', 'the linear distance between centers of chord and circle', or 'the centre of the chord over the interior area of the circle' (1973: 1)? Each corresponds to the different ways of determining the probabilities in Bertrand's paradox (or, in terms of relative entropy, determining the reference prior). If we don't know which parameter to be indifferent towards, we get three different answers, hence the paradox.

Jaynes claimed to have found a way to determine the parameter we should be indifferent towards. His suggestion was that we take inspiration for our search for a reference prior from the way physicists often solve problems: by looking for what he called sameness of problems. We should determine which restatements of a problem should not change our answers to that problem. This leads us to pick out certain invariances, transformations, which should be used to characterize the problem. To help understand what transformations are, imagine that you have a sheet of graph paper, and a small circle cut from red construction paper. Put the circle on the graph paper. Then rotate the circle, without moving its centre from where it is. The circle, if you've cut it well, looks exactly the same. Now move the circle around the graph paper—move it up and to the right. The circle still looks the same. Finally, imagine that the circle and the paper swell up together, at the same rate, and that you step back by an appropriate amount. The circle still looks the same. The circle is therefore *invariant* under rotation, translation, and dilation (dilation invariance is sometimes called dilatation or scale invariance).

Jaynes suggests that the Bertrand paradox can be dissolved by concentrating on finding the correct invariances. We can imagine that the outcome of the experiment of throwing dowels at a circle can also be invariant in these ways: it shouldn't matter where we stand around the circumference when we throw the dowel. It shouldn't matter if the circle moves randomly when the dowel is thrown down (Jarda could have drawn the circle on some paper and moved it around as Prokop dropped the rod. He didn't, however. He drew it in chalk, and since the garage was dark, got some candles to use for lighting. Hence the fright for the Reverend.) Finally, if somehow the circle and the dowel swelled up together it shouldn't make any difference to how often the chord is longer than the side of the inscribed triangle.

Actually, we don't have to imagine that the circle and dowel swell up together. ('Whoa, what the hell was in that beer?') Jaynes uses the idea of observers with different sized eyes. But to make our experiment work given the crude measuring instruments, the eyes would have to be quite different. Perhaps we could recruit a Colossal Squid, *Mesonychoteuthis hamiltoni*, to help with the observations, since they are currently the record holder in the largest-eye category—one of the very few specimens ever caught had eyes an estimated 30cm in diameter when alive. ('Prokop, do you ever feel like you're being watched by a giant cephalopod?')

If we require a solution to Bertrand's problem to be invariant under rotation, translation, and dilation, then the first two answers—indifference between 'the angles of intersections of the chord on the circumference' (probability 2/3) and 'the centre of the chord over the interior area of the circle' (probability 3/4)—are ruled out. They both vary over translations, while the first also varies over changes in scale.

Indifference with respect to the distance from the circumference to the centre is determined as the only possible solution (Jaynes 1973 contains the details, which we shall skip, even in the appendices). We can then apply the Maximum Entropy Principle to determine the probability (density) of a chord falling with a certain length, which gives the answer of 1/2 corresponding to the results of the experiment.

6.2.3 Determination of continuous probabilities

Jaynes seems to have dissolved the paradoxes of geometric probability, and more generally, all such paradoxes generated by the Principle. We are now in a clearer position to see how this has been (putatively) accomplished: the Maximum Entropy Principle should be applied not at the level of events, but at the level of what he terms problems, that is, statements of the situation in which we wish to determine probabilities. Indifference between descriptions of these situations gives rise to a group of transformations, which (may) then determine the probabilities:

> We agree with most other writers on probability theory that it is dangerous to apply this principle at the level of indifference between *events*, because our intuition is a very unreliable guide in such matters, as Bertrand's paradox illustrates.
>
> However, the principle of indifference may, in our view, be applied legitimately at the more abstract level of indifference between *problems*; because that is a matter that is definitely determined by the statement of a problem, independently of our intuition. Every circumstance left unspecified in the statement of a problem defines an invariance property which the solution must have if there is to be any definite solution at all. The transformation group, which expresses these invariances mathematically, imposes definite restrictions on the form of the solution, and in many cases fully determines it.
>
> (Jaynes 1973: 9)

A problem may be described in such a way that the problem would require too many invariances, and so there could be no solution which satisfies them all. Thus a sketchy problem is in fact *overdetermined*, since 'so many things are left unspecified that the invariance group is too large, and no solution can conform to it' (Jaynes 1973: 10). The claim is, then, that Bertrand's problem is neither over- nor underdetermined: the symmetries yield a unique solution—they specify that we should apply the Maximum Entropy Principle in a particular way.

According to Jaynes, the symmetries must be found *first*, and *then* the Maximum Entropy Principle applied. Thus, we could modify von Mises's dictum 'first the collective, then the probability' to express Jaynes's version of the logical interpretation:

> First the symmetry, *then* the probability.

Another way of putting the point is that the Maximum Entropy Principle can be taken as basic, as being fundamental to probabilities, and not vice versa. Hence the Principle does not entail contradictions, since when applied at the level of the determination of probabilities, it either gives one answer or none.

6.3 Maximum Entropy and the Wine-Water Paradox

Jaynes (in 1973) gave up on the wine-water paradox because he thought there was not enough information in the problem to specify any symmetries (it is overdetermined, there are too many solutions, hence the contradictions). Rosenkrantz 1977, however, argued that there *is* a solution. According to him, the problem requires invariance with respect to scale. A fairly straightforward argument (most clearly set out in Milne 1983, but see also the discussion in van Fraassen 1989: 309, 370 n.14) shows that the *log-uniform distribution* satisfies scale invariance.

The log-uniform distribution takes the measure of a quantity falling between two points a and b to be $\log b - \log a$ divided by the difference of the logs of the end points of the scale, where log is the natural log, i.e. base e. We get a familiar-looking formula: for a parameter T that has a range $[a, b]$, the probability that it will fall in the sub-interval $[c, d]$, $p(c \leq \mathrm{T} \leq d)$, is

$$\frac{\log d - \log c}{\log b - \log a}.$$

(This quantity is in fact independent of the base of the logarithm. Also, this version of the Maximum Entropy Principle only works for parameters that range over positive numbers: this can be fixed, see Milne 1983.) Applied to the wine/water paradox, then, we are looking for $p(\text{wine/water} \leq 2)$, which is

$$\frac{\log 2 - \log\frac{1}{3}}{\log 3 - \log\frac{1}{3}} = \frac{\log 2 - \log 3^{-1}}{\log 3 - \log 3^{-1}} = \frac{\log 2 + \log 3}{\log 3 + \log 3} = \frac{\log 6}{\log 9}.$$

For $p(\text{water/wine} \geq 1/2)$ we get

$$\frac{\log 3 - \log\frac{1}{2}}{\log 3 - \log\frac{1}{3}} = \frac{\log 3 + \log 2}{\log 3 + \log 3} = \frac{\log 6}{\log 9}.$$

The contradiction has vanished, and the Maximum Entropy Principle has overcome its greatest challenge.

6.3.1 Problems with the solution—dimensions or not?

Or has it? Milne 1983 points out that there are versions of the paradox which give different answers even when we use the log-uniform distribution.

(Van Fraassen 1989: 307–10, 314 provides a very clear explication of Milne's argument.)

Back in the lab we find a beaker filled with 4cl of liquid; next to it is a note: 'What is the probability that less than or equal to $2\frac{2}{3}$ cl of this liquid is wine? There is at least 1 cl and at most 3 of wine in the glass! I'm drinking your beer next door, Jarda'.

Our calculations should now be routine. The possible range of wine in the drink is 1 to 3 cl, and so $p(\text{wine} \leq 2\frac{2}{3}\text{ cl}) = (\log 2\frac{2}{3} - \log 1)/(\log 3 - \log 1) = \log 2\frac{2}{3}/\log 3$. But, if there is less than $2\frac{2}{3}$ cl of wine in the glass, then the ratio of wine to water is less than $2\frac{2}{3}\text{ cl}/1\frac{1}{3}\text{ cl} = 2$. The contradiction rises again: $\log 2\frac{2}{3}/\log 3 \neq \log 6/\log 9$.

In our original example we used only proportions. In our restatement, we have used measures of volume. The original was dimensionless, the restatement not. As van Fraassen points out, this is enough to resurrect the contradiction, since the log-uniform measure obeys scale invariance, but not translation invariance. But when we use dimensions, we change the invariances needed, and get a contradiction. Perhaps one could argue that the dimensional and dimensionless characterizations are different problems. But they clearly are not. (The debate does not end here: Mikkelson 2004 presents another possible solution; Deakin 2006 argues that it fails.)

6.4 Language Dependence

This need not be a knock-out blow to the Principle of Maximum Entropy. Defenders of the Principle could always return to the position that wine/water type situations are overdetermined, and accept the limitations on the Principle. But the paradox does seem to point to a deeper problem with the Principle of Maximum Entropy—that the probabilities given by the Principle depend on the conceptual and linguistic framework in which the problems are stated. In the next section we examine a famous case that is purported to show that this dependence vitiates any claim to objectivity by the Principle.

6.4.1 The statistical mechanics counterexample

Feller (1957: 39–40) put forward what we will call *the statistical mechanics counterexample*. (This counterexample is retold, albeit with a different moral, in Howson and Urbach 2006: 270–1.) In probabilistic accounts in the physics of the behaviour of elementary particles, the space of possibilities of how a particle behaves (that is, the phase space that describes the possible momentums and positions of a particle), can be divided into, say, n equal regions. We can think of these regions as cells—as small square boxes in a grid. Very small boxes. Suppose we have r particles—think of these as balls. Suppose we don't know anything about how the balls are distributed in the cells. Suppose that we have some particles running around in a definite region, and the particles are not receiving (and have not, for a suitably long period) any outside pushes or pulls. How are the balls distributed among the cells? This is, obviously, an area where arguments based on the Principle of Indifference or of Maximum Entropy can flourish. It seems that we should assign equal probability to each of the n^r possible arrangements (any number of balls can be in a cell). It turns out that assigning equal probabilities allows us to describe how gases behave (using what physicists call for their own reasons Maxwell-Boltzmann statistics—we would call it a distribution). However, no particles actually behave according to Maxwell-Boltzmann statistics.

There is a quantum fly in the ointment: all of the n^r possible arrangements are not allowed. Instead, it depends on the kind of particle. All particles are either fermions (having half-spin, don't worry about what that means) or bosons (having integer spin—don't worry about that either). And fermions and bosons have different symmetries: fermions obey Fermi-Dirac statistics and bosons obey Bose-Einstein statistics. (Exercise: guess who fermions and bosons are named after.) These distributions are flat, subject to certain restrictions (particles can be in a superposition, and so be indistinguishable; no two fermions can be in the same cell): it's just that the possibilities are different.

The moral of the story, we are told, is that there are fermions, and there are bosons, but there are no maxwellions: a priori reasoning has led us astray, and we should stick to looking at the world and experimenting. The Principles of Indifference and Maximum Entropy must simply be wrong, or at least, wrong-headed armchair philosophizing (or probabilizing). And the latter most certainly isn't objective.

6.4.2 Correctly applying the Principle?

The response made by advocates of the Maximum Entropy Principle to the statistical mechanics example is the same as to the wine/water example: the Principle has been misapplied. 'First the symmetries, then maximizing entropy!': wrong symmetries give wrong probabilities. As Rosenkrantz puts it, 'the misplaced faith in Maxwell-Boltzmann statistics was based, not on faulty reasoning or faulty principle of probability, but solely on the mistaken assumption that small particles are physically distinguishable. (It is no criticism of the maximum entropy rule that it leads to incorrect results when you feed it misinformation.)' (1977: 60).

Jaynes also points out that there's more to the counterexample than is usually told: Maxwell was in fact able to make surprising predictions in kinematics, and so the Principle did work (for a different problem):

> It is, however, a matter of record that over a century ago, without benefit of any frequency data on positions and velocity of molecules, James Clark Maxwell was able to predict all these quantities correctly by a 'pure thought' probability analysis which amounted to recognizing the 'equally possible' cases. In the case of viscosity the predicted dependence on density appeared at first to contradict common sense, casting doubt on Maxwell's analysis. But when the experiments were performed they confirmed Maxwell's prediction, leading to the first great triumph of kinetic theory. These are solid, positive accomplishments; and they cannot be made to appear otherwise merely be deploring his use of the principle of indifference.
>
> (Jaynes 1973: 7)

The response to the statistical mechanics counterexample is, as usual, therefore, that when correctly applied, the Principle gives the correct answer, and when incorrectly applied, it does not.

This raises the problem of determining when to apply the Principle, that is, of finding a further Principle to determine when it's being correctly applied. In other words, we need an account of induction. Hence, Rosenkrantz and Jaynes seem to take the position that the Maximum Entropy Principle is not a theory of induction, and we are therefore still in need of one. For example, in the continuous case we must first find the correct prior. But the theory does not tell us how to find this prior, only that it must be determined by the correct symmetries. We seem to be left with the maxim 'Use symmetry when you should use symmetry'. But this we already knew. (This echoes Venn's criticisms of the Rule of Succession in 5.2.1.)

In defence of the Principle, it must also be pointed out that both relative frequency and Bayesian accounts of probability are also dependent on having the correct input, as was pointed out for example in sections 1.2.6 and 3.8. (This point is also made by Rosenkrantz 1977: 60.) That being said, the Principle is quite restrictive in that it requires that everyone adopt a flat distribution. Neither the relative frequency nor Bayesian interpretations require this, and so do not require that everyone adopt the same possibly mistaken degrees of beliefs.

It should also be pointed out that since many problems present us with a vast amount of possible symmetries, locating the correct one is a daunting task. For example, Marinoff 1994 criticizes Jaynes's solution to Bertrand's problem, arguing that his experimental evidence does not confirm the solution he thought it does. (Marinoff 1994 also contains an incisive account of possible problems, associated symmetries, and solutions associated with Bertrand's problem.) It's also worth noting that in the case of relative entropy in 6.2.1 we need a distribution $q(x)$ to serve as the flat distribution, which then is used to measure the divergence of the other distribution $p(x)$, and so give a measure of entropy. But the entropy is truly *relative* to the distribution $q(x)$, and so is 'objective' only to the degree that the choice of the reference distribution is.

The status of symmetry in arguments in physics has been the subject of philosophical curiosity, since it seems quite odd indeed that we should be able to determine without empirical evidence the correct structure of reality via symmetry arguments. And indeed, this seems not to be the case. It could be argued instead that we look for empirical evidence that there are certain symmetries in nature (Kosso 2000 addresses this issue). This highlights another difficulty, put quite nicely by Strevens 2005b, namely, that Principles of Indifference are ambiguous between requirements of *epistemic* and *physical* symmetries. The former would seem to be the correct field of application, but this is not supported by appeal to the latter. Jaynes's claim of support for the Principle from Maxwell's successful prediction confuses these two. There are certainly symmetries in nature, and science is often concerned with these. But this does not imply that the correct inductive method should be the symmetric distribution of probability over all possibilities. (Strevens also makes a telling point by applying the Bookmark paradox to theories of dark matter; do we count competing general programmes like those that postulate new sorts of particles, or do we count the multitude of theories they encompass? The two different counts give different answers, and both seem wrong. It just seems wrong-headed to portion out support on the basis of how many candidates there are for the correct general programme or theory.)

6.4.3 Language dependence and ontological dependence

Probabilities given by the Maximum Entropy Principle depend, then, on what we take the symmetries to be. These symmetries will depend on our language, and, more generally, our conceptual framework. The clearest and simplest example of this is the Maximum Entropy Principle's vulnerability to the Bookmark paradox: the probabilities prescribed by the Principle change depending on how finely we divide up possibilities. And there is no easy rule we can apply to determine the symmetries. For example, we might say that the partition should be as fine as possible. But this would give the wrong answer, since, for example, the partition determining symmetries for fermions is not the finest available—it was in this context that Howson and Urbach employed the statistical mechanics counterexample.

It is not merely the dependence on a language that is problematic, but the dependence on the ontology of that language as well. The reason the statistical mechanics example comes about is that 'particle' may look like the same term in both classical and quantum mechanics, but in fact it's quite a different beast, since in one particles are distinguishable, and in the other not. A similar example is to be found in the change from classical to relativistic mechanics, where the ontology of 'mass' underwent a profound change. Finally, Howson and Urbach consider the Bertrand paradox from the point of view of the theory of special relativity: Jaynes's solution will not be the same if we take into account the frames of reference of the experimenter and the circle (Howson and Urbach 2006: 284, see also Marinoff 1994).

Changes in ontology are prior to the application of the Principle, since the Principle is not the foundation of a theory of inductive learning. This means that the Principle's probability assignments are based on a given language, and then on a given conceptual framework. It seems odd, given these dependencies, to claim that the Principle necessarily leads us to the correct answer. In particular, it seems peculiar to term the interpretation of probability based on the Principle 'Objective Bayesianism'.

6.4.4 The scope of the Maximum Entropy Principle

The insistence on symmetries means that the scope of the Maximum Entropy Principle is quite narrow. There seem to be cases where we do not have relevant symmetries, but are nonetheless willing to assign (reasonable-seeming) probabilities. For example, it seems that the chance of Liberia going to war with Germany in the next year is vanishingly small. If we used only the Maximum Entropy Principle, the problem would be underdetermined (or would require the construction of a rather complex hypothesis to which symmetry could be applied, and then gathering evidence for and against the hypothesis). Therefore, the Maximum Entropy Principle can be seen as a limited case of Bayesian inference, one in which relevant symmetries do determine a flat distribution and are reflected in degrees of belief by a flat probability distribution. This means, however, that the choice of the Maximum Entropy Principle is a Bayesian choice. Yet this seems to be the justification often given by proponents of the Principle themselves. Naturally, this conclusion has been resisted by the proponents. We turn to another means of justifying the Principle in the next section.

6.5 Justifying the Maximum Entropy Principle as a Logical Constraint

Part of the difficulty in justifying the Maximum Entropy Principle as an inductive method has been the pragmatic way in which we have approached the Principle. So far, we have not discussed arguments for the justification of the Maximum Entropy Principle, appealing merely to its intuitiveness. This carries a whiff of ad-hocness: why $1/(p(A))$, and not another function? One answer is, of course, that the whiff is not of ad-hocness, but of post-hocness. This definition has, as we have seen, many desirable properties, which seem to overcome limitations with the traditional Principle of Indifference. But these properties alone don't give us a logic of induction. However, if we could demonstrate that there were strong a priori reasons for adopting the Maximum Entropy Principle, it might go some way towards answering the criticisms of the preceding sections.

There are, however, mathematical results that are claimed to provide a priori justifications for the Maximum Entropy Principle by establishing that entropy is the only function that satisfies certain criteria.

6.5.1 Maximum entropy as imposing consistency

The first proof that entropy is the only function to satisfy certain criteria was provided by Shannon (1948: 10–12). Since Shannon aimed only to establish a theory of signalling and communication, as we have mentioned earlier, it fell to Jaynes 1957a to make the link with the idea of an inductive logic.

In the continuous case, a related justification was put forward by Shore and Johnson 1980. There are also a number of recent papers by Paris and his collaborators (Paris 1994, Paris and Vencovská 2001) that explicitly appeal to (mathematical and philosophical) logic. The arguments of Shannon, Jaynes, and Shore and Johnson, among others, are critically discussed in

Uffink 1996, a classic in the field. Howson and Urbach 2006: 286–7 briefly discuss Paris and Vencovská 2001. We will only discuss the Shannon/Jaynes justification, since the others are well beyond the scope of this book. The points we will make about it may be applied to the other justifications as well, however.

In section 6 of his 1948 paper, Shannon provided (an outline of) a theorem that says that the entropy function is the only function that satisfies three desiderata, which some find compelling. In other words, he showed that there is only one function H of probabilities such that:

1. H is continuous with respect to the probabilities it is defined over (a small change in the probabilities means a small change in H).
2. When the probabilities are equal, H is a monotonically increasing function of the number of alternative possibilities (H is greater when there are more values of random variables, and hence more values of the associated probability functions/random variables).
3. The quantity H associated with a series of random variables can be represented as a weighted mixture of subsequences of the series of random variables:

$$H(p(A_1),\ldots,p(A_n)) = H(p(A_1)+\ldots+p(A_m), p(A_{m+1})+\ldots+p(A_n))$$
$$+ (p(A_1)+\ldots+p(A_m))H\left(\frac{p(A_1)}{\sum_{i=1}^{m} p(A_i)},\ldots,\frac{p(A_m)}{\sum_{i=1}^{m} p(A_i)}\right)$$
$$+ (p(A_{m+1})+\ldots+p(A_n))\,H\left(\frac{p(A_{m+1})}{\sum_{i=m+1}^{n} p(A_i)},\ldots,\frac{p(A_n)}{\sum_{i=m+1}^{n} p(A_i)}\right)$$

The last axiom calls for an illustration, which we can find in Prokop's kitchen. Prokop has a particular cabinet (illustrated in Figure 6.4) devoted to some base ingredients that determine what's for dinner. Right now there's not so much variety. There are three compartments: the rice compartment; the noodles compartment; and the panko compartment. Given the ingredients gathered he can make a fish head curry (rice), a mie goring (noodles), or he can (and Jarda fervently hopes he does) fry some fish (panko).

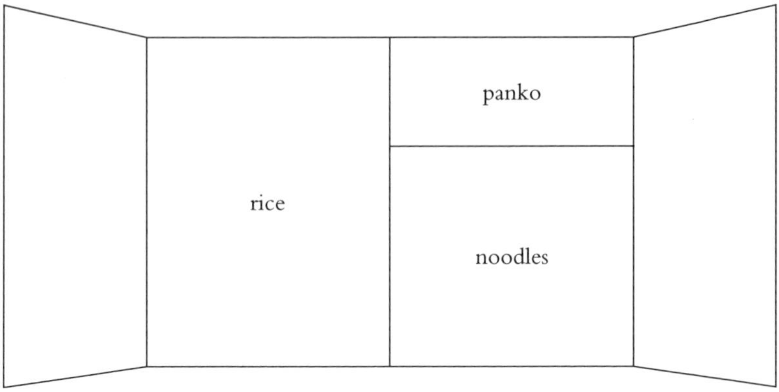

Figure 6.4 Prokop's dry carbohydrates cabinet

But how to choose? He needs some randomness in his life. A blindfolded Jarda will serve well: the choice of menu will be determined by what Jarda randomly draws out (no, Jarda does not go in the kitchen much). With some idealization (well, a lot, actually), we can imagine that Jarda's chance of drawing a particular food is based solely on the percentage of volume that food's compartment takes up of the total. As you can see, the rice takes up 1/2 of the total, egg noodles 1/3, and panko 1/6.

Now, suppose Prokop guides Jarda to the cabinet, but has forgotten to open its doors. If Jarda opens the left door, fish head curry it will be, but if he opens the right, there is now a 2/3 chance of egg noodles and a 1/3 chance of panko. Jarda is perfectly ambidextrous, and so there's a 1/2 chance of either door being chosen.

The grouping axiom relates these two situations: the second is simply a rewriting of the first, since the ultimate probabilities are the same. So the first stage (opening the left or right door) has entropy $H(1/2, 1/2)$, and the second stage has $H(2/3, 1/3)$. But since it will only happen half of the time, we give it the weight of 1/2, that is:

$$H(1/2, 1/3, 1/6) = H(1/2, 1/2) + 1/2\, H(2/3, 1/3)$$

This says that, from the point of view of entropy, the second situation (doors closed) is the same as the initial situation with doors open. More generally, if the outcomes have the same probability, the order of their choice is irrelevant.

The claim is that these properties capture the notion of uncertainty necessary to ground an (objective) interpretation of probability on the notion of consistency alone. But an examination of the axioms shows that this is not, at least straightforwardly, the case. For example, the first property is 'nice': it's usually explained, somewhat problematically, as 'small variations in probabilities lead to small variations in entropy'. The second axiom says that greater numbers of equally likely alternatives increase uncertainty. The third axiom can be taken as saying that entropy is independent of how the probabilities are given: either in groups or separately. It is thus a sort of symmetry condition, and so it's perhaps not surprising that the three axioms give us the Maximum Entropy Principle.

6.5.2 Problems with the Maximum Entropy Principle as consistency

As Uffink (1996: 231–3) points out, the axioms do not seem to be about consistency. There are no references to contradictions, for example, or about extendability of a model as, for example, in Howson's interpretation in 3.4.3. That is not to say that such an interpretation could not be offered —but thus far none has been. If we take the axioms as conventions, then these conventions should be compelling: they should capture undeniably intuitive properties of information. There is reason to think that they do not.

Seidenfeld (1979: 421–3) provides an example of how the requirement that the order of data not affect uncertainty violates conditionalization, for example. So much the worse for conditionalization, a proponent of the Principle may respond. But the difficulty is quite general. The relation of conditional entropy, the entropy for A given that B has occurred, is defined as

$$H(A|B) = -\sum_{n} p(A_n) \sum_{m} p(A_n|B_m) \log p(A_n|B_m).$$

It is then possible to prove that $H(A|B) \leq H(A)$ (see Ash 1965: 239). This means that learning new information never increases entropy/uncertainty. This does seem reasonable at face value: if the information in B has nothing to do with A, then adding it does not decrease total uncertainty. But if learning B is learning something new, then it should decrease the amount of uncertainty associated with A. A bit more thought shows, however, that it depends very much on what is learned: many data learned increase overall uncertainty, particularly when they show that we were quite wrong before about something. An example from Uffink (1996: 234 and 1990:

72–3): I am quite sure about the location of my keys. But then I learn, heart-stoppingly, that they are not as I thought in my pocket. My overall uncertainty about the location of my keys dramatically increases (and even more so if I am travelling). Uffink (1990, sections 1.6.2–1.6.3) also subjects the third axiom to a penetrating analysis and critique.

Uffink (1996: 234) notes that Jaynes had considered such counterexamples. He himself uncovered what he took to be a paradox: a mathematical approximation of a physical process has less entropy, and hence, for Jaynes, more information, than the exact account. He then continues:

> This paradox shows that 'information' is an unfortunate choice of word to describe entropy expressions. Furthermore, one can easily invent situations where acquisition of a new piece of information (that an event previously considered improbable had in fact occurred) can cause an increase in the entropy. The terms 'uncertainty' or 'apparent uncertainty' come closer to carrying the right connotations.
>
> (Jaynes 1957b: 186)

Uffink responds that this is 'rather disappointing', since the Maximum Entropy Principle has been now relegated to a measure of 'apparent uncertainty', which hardly seems to bear the weight of the term 'objective', much less a consistency constraint.

There are also questions about whether this measure of information, and in particular the grouping axiom, is appropriate in a quantum setting: Brukner and Zeilinger 2001 argue that it is not; Timpson 2003, Mana 2004, and Hall 2000 argue that, appropriately interpreted, it is. This is exactly where one would expect controversy to arise, since one would not expect invariance with respect to how groups of data are received in a quantum setting.

There is also considerable discussion as to whether the notion of information is really our ordinary one. It is clear when reading Shannon that he used the term only as a measure of successful coding: he scrupulously avoided any broader application of the term. Later followers have not been so circumspect. Hayles 1999 is a good starting point for discussions of the uses as well as abuses of the notion of entropy and information in philosophy and elsewhere.

6.6 Conclusion

The logical interpretation, as we have seen, lives on—if anything it seems to be growing in popularity. As we saw in the first part of this chapter, there is good reason: the method is elegant, and gives a clear formulation of the Principle of Indifference. Yet despite this, it is still subject to the same problems as the older versions. It is dependent on language. The continuous case is highly problematic. These two difficulties should be enough to show that the Principle will not give us an objective account of inductive reasoning. Jaynes has done a great service in clarifying when the Principle can be applied in the continuous case non-paradoxically. In particular, his advice to specify the necessary symmetries beforehand is an important caution. And yet we are still left without a means for choosing the *correct* symmetry, and so are no further along to an object theory of inductive inference. So it seems that the Maximum Entropy Principle is best viewed as one of the tools in the Bayesian kit: not a foundational method, but a part of a larger, subjectivist method. However, for a rather different conclusion, see Williamson 2010.

Appendices

These appendices are a mixed bag: they cover some basic mathematics, and a few topics I think should be covered in a book like this, but which don't fit in the main narrative. While the appendices are optional, I hope they may provide at least some diversion and instruction. However, the appendices are meant as summaries of what you might want to know to continue the study of the foundations of probability, and so should be taken more as pointers than as introductions to unknown fields.

A.0 Some Basics

A.0.1 Percentages

A percentage (*per cent*, i.e. part of 100) is a number. It's meant to represent the division of an amount into parts of 100. So, 1 per cent is simply 1/100th of something, 10 per cent is 10/100th, or 1/10th, and, of course 100 per cent is all of something. An example: someone tells you that 10 per cent of the people in a 200-hundred-person crowd are badly dressed. This means that 10 out of each 100 persons in the crowd are badly dressed, that is 20 of them are committing crimes against fashion. So, to determine a percentage of a number, we multiply the number by the x, where x is the percentage.

A.0.2 Kinds of numbers

There are many different kinds of numbers. The most familiar, perhaps, are the positive integers, also referred to as the counting numbers: 1, 2, 3, . . . Rational numbers are numbers that can be represented as ratios of integers: 1/2, 22/7, 5/15 . . . Irrational numbers are numbers that are not rational, that cannot be represented as ratios of integers: $\pi, \sqrt{2}$. Real numbers are the rationals and the irrationals. (There are, of course, many other types of numbers such as transcendental, algebraic, odd, even, negative, and complex.)

A.0.3 Sizes of sets—countable and uncountable

To the continuing surprise of generations of students, there are different sizes of infinity. The (sets of the) integers and the rationals (numbers got by dividing integers) are the same size, and are smaller than the set of the real numbers. This was first shown by Cantor. We say that two sets are the same size if we can match up all their members in exclusive pairs, that is, if we can match up each member of one set with exactly one member of the other set. If we can match up a set with the counting numbers, the positive integers, we say the set is *countable*. If not, the set is *uncountable*.

A.0.4 Functions, limits

I was tempted to follow oral tradition and describe a function as taking some object, known as the *argument*, and either returning an object, or nothing at all. This, of course, won't do unless we know what 'taking' and 'returning' mean: is the grocer a function when he takes your money and gives you an apple, or runs away with it, giggling hysterically? The grocer may not be a function, but we can abstractly describe the situation as a link between your coin and the apple, or no apple, as the case may be. So it's better to think of a function as a mapping from a domain of objects to a range of objects, where at most one member of the range is assigned to a member of the domain. (Alternatively: as a set of pairs of objects.) The important thing is that in no case is more than one object assigned to a member of the domain.

The notation $\lim_{n\to\infty} f(x) = c$ means that as x grows arbitrarily large $f(x)$ approaches c. After some point, $f(x)$ will remain close to c in the sense that for every real number $\varepsilon > 0$, there is a natural number $\delta > 0$, such that for $x > \delta, |f(x) - c| < \varepsilon$.

A.0.5 Logarithms

$\log_a x$ is the number which you need to exponentiate a to get x. In other words,

$$\log_a x = y \text{ iff } x = a^y$$

a is called the base of the logarithm. If the base of the logarithm is the number e, then the logarithm is called the *natural* logarithm, and is sometimes denoted as *ln*. e is an interesting number (Maor 1994 covers its history and uses) but for our purposes we encounter e because, like logarithms, it often shows up when dealing with exponential quantities. ($e = \lim_{n\to\infty} (1 + 1/n)^n$.) We have almost always used base 2, since it helps us express bits, except for when we used the natural logarithm in section 6.3. Another common base is 10, hence the name 'the *common* logarithm'.

After a bit of practice, working with logarithms becomes quite natural. The following three equations are essential to the understanding of logarithms.

$$\log_a xy = \log_a x + \log_a y,$$

$$\log_a\left(\frac{x}{y}\right) = \log_a x - \log_a y$$

$$\log_a x^n = n\log_a x.$$

It should also be noted that

$$\log_a a = 1$$

and

$$\log_a 1 = 0.$$

The standard convention is that $0 \log 0 = 0$.

A.1 The Axioms

The axioms of the probability calculus can (sometimes) be nicely visualized as percentages using Venn diagrams.

Let us consider the percentage the set A takes up of the whole space we are measuring, $p(A)$. Let us call this whole space, as it usually is in discussions of probability, Ω. By convention we set the size of the whole space to 1, that is,

(1) $p(\Omega) = 1.$

Of course, the percentage of the space any set takes up, if any, will always have to be positive, so for any subset A,

(2) $p(A) \geq 0.$

Now suppose we have two different sets, A and B, that don't overlap. Then, the percentage of the space they take up will add. That is,

(3) $p(A \cup B) = p(A) + p(B)$, when $A \cap B = \varnothing$.

The corresponding Venn diagram for this is

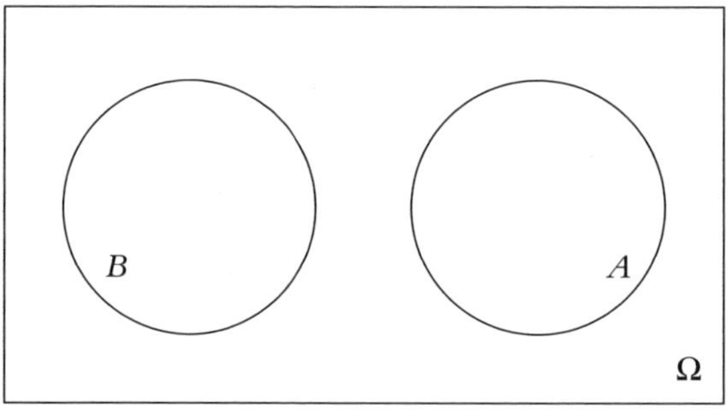

It is left for the reader to work out what happens if the sets overlap, as in the following figure.

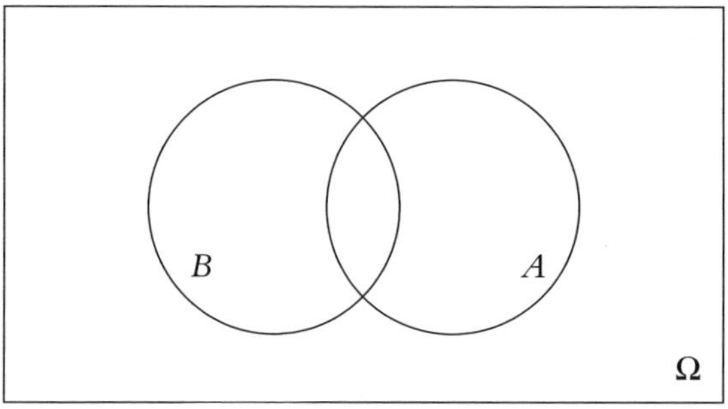

We can also use the diagrams to see that the *complement* of A, that is $\Omega - A$, has probability $1 - p(A)$.

The reader should be wary: not all statements about probability are easily visualizable, and we should not rely too much on diagrams.

A.1.1 Conditional probability, independence

There are two other notions which are of particular importance for probability theory: conditional probability and independence.

Suppose we have two events, A and B. The percentage of B events that are also A events is known as the percentage, or the probability, of *A conditional on B*. (So, if 10 per cent of B events are A events, then the probability of A conditional on B is .1.) Another way of looking at it is that first we only look at B events, and then check to see how many of these are A events. In other words, we look for how many A's are B's. This probability is written $p(A \mid B)$, and is $p(A \cap B)/p(B)$, given that $p(B) \neq 0$.

The other notion of special importance is independence. Two events are said to be *independent* of one another if the probability of their happening together is equal to their individual probabilities, that is,

$$p(A \cap B) = p(A)p(B).$$

Another way of putting it is that two events are independent if the probability of one conditional on the other is just its probability, that is,

$$p(A \mid B) = p(A).$$

This might be read as saying that the occurrence or non-occurrence of B has no effect on the probability of A.

A.2 Measures, Probability Measures

Probability theory is a part of measure theory (which itself is a part of analysis). Measure theory is the aptly if unimaginatively named abstract study of measures. Length, width, breath, depth, and area are all familiar measures. Measure is, I suppose, at bottom trying to find out how many things fit into another thing: how long is the length of this foot? If we can fit 11 inches into that length, it's at least 11 inches long. How much of the price of this beer is taken up by tax? We divide the cost of beer into some standard units, and then see how many match up with the tax. Thus measure can be seen as a fraction: a meter is 1/10,000,000 of the distance from the North (or South) pole to the equator, for example. Thus a meter is a percentage of another length.

Mathematically, a measure is a function over sets that assigns 0 to the empty set (the size of nothing is 0) and adds over disjoint sets (if each of your feet is 1 foot long, the length of your two feet together is 2 feet). Probability is, like percentage, a *normalized* measure: it is a measure, but instead of ranging from 0 to ∞ like length, the end points of its range are finite—in this case 0 and 1.

A.2.1 Fields

We still have not said much about exactly which sets probability should be defined over. Obviously, we would like to talk about all possible combinations of sets. If we have sets A, B, C, D ... and so on, we would like to be able to talk about all their possible combinations, $A - C$, $B \cap D$,

$(A \cap B) \cup D^c \cap \ldots$, and so on. The nicest way to do this would be to define probabilities for all sets of subsets of Ω. It turns out that this cannot be done when dealing with the infinite combinations discussed in the next appendix, which means that there are non-measurable sets. We will not have need to discuss them here; I merely mention them because they are cool.

Still, we want to define probability over as many sets as possible. To do this, we need some jargon. The reader is warned that the jargon in this appendix is meant to be nothing other than abstract labels, for now devoid of meaning. We start with our basic set, Ω, which is usually called the *sample space*. (Don't be misled—Ω is just a set which may or may not have anything to do with 'sampling'.) The elements in Ω are termed *elementary events*. Subsets of Ω are *events*. (Don't think philosophically—the elementary events are points in the set, events are collections of those points.)

The collection of subsets of Ω that will take probabilities is called a *field*, or a *field of sets*, and is denoted F. It is defined as follows: first, if there is some event, call it A, that is a member of Ω, then its complement (A^c or, in some contexts, $\neg A$) is also in Ω; second, if two events A and B are in F, so is their union, that is, the event of either one or the other (or both) of them happening; finally, we want to be able to talk about the whole set Ω (*the certain event*, i.e. that *something* happens), and we want to be able to talk about the empty set (the impossible event, i.e. that *nothing* happens), and so they are in the collection too. Any collection of subsets of Ω formed in this way is a field.

Probability is then defined over a field of sets (of some set). The probability function together with the set Ω and the field of sets over Ω is called a *probability space*. A formal definition of a field can be found in Appendix A.2.1.

This characterization of the probability calculus was put into its (mostly final) form by the mighty Russian mathematician Andrei Nikolaevich Kolmogorov (Андрей Николаевич Колмогоров). Although probability had long been studied, Kolmogorov provided an abstract characterization of the probability calculus that fully satisfies mathematicians. It is also simple, elegant, and flexible enough to clarify discussion of different interpretations of the probability calculus.

Now we get on to some more substantial, but still quite simple, mathematics—if you relax. A *set function* maps sets to real numbers, that is, it's a function from sets to real numbers. We will be interested in certain collections of sets, fields.

A.2.2 Fields, σ-fields

More formally, a field F of subsets of Ω is any collection of subsets that contains the empty set and Ω, and if any two subsets of Ω, A and B, are in F, then so is their intersection and union, as well as their complements. This implies that for any finite number of subsets of F, their intersection is in F. If we extend this to infinite sets, that is, for any countable group of sets their intersection is in F, then we have a σ-field. Even more formally, let F be a collection of subsets of Ω such that

(i) $\Omega\ F, \varnothing \in F.$

(ii) if $A_i \in F, i = 1, \ldots, n$ then $\bigcup_{i=1}^{n} A_i \in F.$

(iii) if $A \in F$, then $A^c \in F.$

Then F is a field (also called an algebra).

If we allow infinite unions, that is, if we replace (ii) by

(ii′) if $A_i \in F, i = 1, \ldots, n$ then $\bigcup_{i=1}^{\infty} A_i \in F.$

then F is a σ-field (or σ-algebra). (The smallest σ-field consists of Ω and $\varnothing$, as you can now prove.)

Fields ensure that we can describe events and collections of events. Suppose our basic events are two coin tosses. Then the set of all possible outcomes of the tosses is HH, HT, TH, TT. We can combine these *basic* events to create other, *compound*, events. For example, the compound event HH ∪ HT ∪ TH is the event of heads occurring at least once; HT ∪ TH is heads occurring exactly once. The (set-theoretic) complement of HH ∪ HT ∪ TH with respect to the entire set of outcomes is TT, the event of heads never coming up, for example.

A.2.3 Measures

Now we wish to assign measures to the sets in the field. There are two obvious restrictions: first, the measure of nothing is nothing, that is, zero. Secondly, a measure should be additive, that is, the measure of two non-overlapping sets should add. More formally,

$$\mu(\varnothing) = 0$$

$$\mu(A \cup B) = \mu(A) + \mu(B), \quad A \cap B = \varnothing, A, B \in F.$$

A.2.3.1 Measure zero Consider the size of a point relative to a line. The point is infinitely small, and so has measure 0. Or consider a line in a plane. Since a line has no width, it has measure 0 with respect to the plane. Similarly, countable sets are infinitely small with respect to the continuum. They form sets of measure 0. Since measure theory is abstract, many sets can have measure 0, be they bets or sequences of attributes.

A.2.4 Probability measures

Probability measures are measures from a (σ-)field generated by a sample space $\boldsymbol{\Omega}$ to the interval between 0 and 1. (In what follows we will denote members of the field as $A, B, A_1, \ldots, A_n$.)

(1) $\quad p(A) \geq 0 \quad \text{for all } A \in \Omega,$

since measures are 0 or positive.

(2) $\quad p(\Omega) = 1,$

since the probability measure can't be greater than 1.

(3) $\quad p(A \cup B) = p(A) + p(B), A \cap B = \varnothing.$

And, of course, the measures add.

Generally, for $A_i \cap A_j = \varnothing$

(3a) $$p\left(\bigcup_{i=1}^{n} A_i\right) = \sum_{i=1}^{n} p(A_i)$$

3a is the property of *finite* additivity. Textbooks usually postulate the stronger property of *countable* additivity:

(3b) $$p\left(\bigcup_{i=1}^{\infty} A_i\right) = \sum_{i=1}^{\infty} p(A_i)$$

The triple $\langle \Omega, F, p \rangle$ is known as a *probability space*.

Sometimes the following is added as an axiom:

(4) $$p(A \mid B) = \frac{p(A \cap B)}{p(B)}, p(B) \neq 0.$$

However, this can also be taken as a definition (or not: see Hájek 2003).

A.2.4.1 The philosophical status of countable additivity Countable additivity has hardly been taken as self-evident. Instead, it seems usually to be conceived of as a convenience. One way of seeing this convenience is to restate countable additivity as a continuity condition. Consider a chain of members of a σ-field $A_1 \supset A_2 \supset \ldots \supset A_n \ldots$ Let A be the smallest set in this sequence, i.e. $\bigcap_n A_n$. A natural continuity condition would be that

If $\bigcap_n A_n = \varnothing$, then $p(A_n) = 0$.

However, this does not follow from the axioms for finite probability if the chain is infinite, and so an additional requirement is needed, that is

If $\lim_{n\to\infty} \bigcap_n A_n = \varnothing$, $\lim_{n\to\infty} p(A_n) = 0$

(see e.g. Billingsley 1995: 25). Kolmogorov 1933: 15 takes this as a matter of convenience, of idealizing away from actual physical processes, and when countable additivity is seen as a continuity condition, the mathematical convenience is obvious. But it also carries philosophical baggage.

For relative frequency interpretations, the assumption of countable additivity can raise difficulties of the type mentioned in section 1.3.1, since relative frequencies are not in general countably additive. A standard argument, which van Fraassen attributes to Birkhoff, can be found in Giere 1976, van Fraassen 1980, Gillies 2000, Howson and Urbach 2006 (as well as in the earlier editions), and Howson 2008. The basic idea is simple. Assume a sample space with countably many basic events, where each event occurs only once. Then the limiting relative frequency of any event is 0. But the union of the events should have probability 1. Since the countable sum of 0s is 0, not 1, relative frequencies are not countably additive. (Gillies uses the example of engines receiving unique serial numbers; van Fraassen uses the occurrence of a particular day.)

Humphreys (1982: 141) has pointed out that, contrary to popular opinion, this objection applies neither to von Mises's interpretation nor to the Kolmogorovian interpretations of 1.3.3 and 1.3.4. However, countable additivity does raise problems for anyone with a very restrictive empiricist outlook. As Kolmogorov put it, 'Since the new axiom [countable additivity] is essential for infinite fields of probability only, it is almost impossible to elucidate its empirical meaning . . . For, in describing any observable random process we can obtain only finite fields of probability. Infinite fields of probability occur only as idealized models of real random processes' (1933: 15). Von Mises is usually taken to agree with this account, and so restrict his interpretation to countable additivity. Joint work with Peter Milne has convinced me that this is not in fact the case, but this is not the place to discuss it. Instead, it suffices to note that a lack of countable additivity is often taken as a reason to reject an interpretation of probability out of hand.

De Finetti (1974: 120 and 1972, chapter 5) thought that countable additivity was troublesome for a subjective interpretation of probability. A standard example is an infinite fair lottery, that is, a lottery with infinite tickets, each indexed by a natural number, each with an equal chance of being drawn. A natural thought is that since each ticket has a chance of being drawn, and since the lottery is fair, we should assign some very small probability to the chance of each ticket being drawn. But this violates countable additivity: the probability of some ticket winning is 1 by the first axiom; but by countable additivity, the probability is infinite, since we infinitely sum positive probabilities; hence, a contradiction. Hájek 2003 has a nice discussion; Howson 2008 is a recent defence of finite additivity.

A.2.5 Some useful theorems

$$p(A \cup A^c) = p(A) + p(A^c) = 1$$

$$p(A) = 1 - p(A^c)$$

$$p(A|B) = \frac{p(B|A)p(A)}{p(B)}, \text{if } p(B) \neq 0.$$

This last is known as Bayes's theorem. The denominator is usually calculated using the theorem of total probability,

$$p(A) = p(A|B)p(B) + p(A|B^c)p(B^c)$$

or, more generally,

$$p(A) = \textstyle\sum_i p(A|B_i)p(B_i)$$

where the B_i partition Ω, that is, their union is Ω and their intersections are the null set $\varnothing$.

A.3 Random Variables

A random variable is a function that has as its domain the sample space and as its range the real numbers. (Do not be misled by the unfortunate name: a random variable is not a variable, but a function!) It is often more convenient to work with random variables than the underlying sample space, since we then work with numbers, and not labels.

The usefulness of random variables can be seen in the simplest non-trivial example, the Bernoulli random variable, which takes only two values, 0 and 1. We can use this to represent the outcome of a coin-tossing experiment (i.e. a sample space of {heads, tails}).

$$X = \begin{cases} 0 \text{ if tails} \\ 1 \text{ if heads} \end{cases}$$

In this case, $p(X = 1)$, that is, the probability of heads is $1 - p(X = 0)$, as can easily be proved. (If we were being exact, we would write $X(\text{heads}) = 1$, $X(\text{tails}) = 0$. But we follow tradition in not being exact, and omit reference to the sample space, writing only $X = 1$, $X = 0$, etc. Still, it is important to keep in mind that X is a function from the sample space, that is, possible outcomes of an experiment, to numerical values.)

A.3.1 Sums of random variables

We can sum random variables in the obvious way. Suppose we have a series of coin tosses. Then we can define the function X_n that takes the value 0 if the nth toss results in tails, and 1 if heads. Then $X_1 + X_2 + X_3$ is just the first three tosses added (if they were all heads, it would be 3, if all tails 0, etc., two heads 2, etc.). More generally, the sum of the first n random variables is

$$\sum_{i=1}^{n} X_i,$$

which, obviously, tells us the sum of the values of the n random variables, which in turn tell us the number of positive outcomes in n trials. It is often convenient to denote the sum of n random variables as S_n.

Random variables allow us to express many ideas simply. For example, the relative frequency of heads in n tosses is the sum of the random variables S_n divided by n. The limiting relative frequency of heads is

$$\lim_{n \to \infty} \frac{S_n}{n}.$$

Let us take the more complicated example of rolling a die. In this case the random variable, let's call it A, has six outcomes. The *distribution function* of a random variable is

$$p(A \leq x).$$

Suppose, for example, that the six outcomes of a roll are equally probable. Then $p(A \leq 4) = 4/6 = 2/3$. The *probability mass function* is

$$f(x) = p(A = x).$$

The distribution function and the mass function are interdefinable, but we're not going to interdefine them here.

A.3.2 Expectation

When we want to know the average grade a student earns in a class, we add up the grades and divide by the number of grades given. This information can throw light on, for example, if the teacher is an easy grader or not. We can also compute the mean of a probability distribution, which will tell us the most likely outcome.

The mean of a random variable is called the expectation. (The name is obvious—it tells you where the probability is concentrated. But it is also misleading, since expectation has many other different meanings.) Expectation is defined as

$$E(X) = \sum_{f(X)>0} xf(x),$$

where $f(x)$ is the probability mass function of the random variable X which takes values x. When convenient, we shall refer to the mean as μ.

A.3.3 Continuous random variables

So far we have only dealt with *discrete* random variables: variables taking at most countably many values. *Continuous* random variables take continuum many values. In the continuous case we work with a *probability density function*, that is, a function f such that $\int_a^b f(x)dx$ gives the probability of x being in the interval $[a, b]$. The expectation is $\int_a^b xf(x)dx$.

A.4 Combinatorics

Prokop is arranging a sampling evening of his Pilsen-style lagers (that's real Pilsen-style, as in, the kind of beer brewed in Pilsen 20 or so years ago). If he were to offer (stingily) three different brews, in how many different orderings could he offer samples of beer? Pilsen Amber, Homesick Pilsener Brew, and Pilsen Cadillac are the choices, which we'll call A, B, and C. We *could* work the answer out by brute force: ABC, ACB, BAC, BCA, CAB, CBA. But that would be a drag. And would leave room for doubt: are we sure we got all the orderings possible? We can't be sure until we've got a more abstract grasp on how many orderings there could be.

Luckily, it's not that hard to work out how many different orderings there can be. Imagine the choice as a tree: at the first stage we have three branches, our first starting beer. Then, at the second stage, we have two branches, the remaining two beers. The third stage has only one branch, since now there's only one beer to choose from. So, there are $3 \times 2 \times 1$ different (drinking) paths.

We can generalize this way of thinking to find out how many ways there are of ordering any set of objects. Suppose we have n objects. Then, there are n ways the sequence can start. But then having chosen one way, there are $n - 1$ ways the sequence can continue. And so on: $n(n-1)(n-2)(n-3)\ldots$ This is abbreviated as $n!$ (and pronounced n-factorial).

A.4.1 Permutations

Now suppose that Prokop is holding a tasting of his beers, and he's not going to be stingy. But he doesn't want his friends to get horribly drunk and destroy the vines next to the apartment by swinging from them, shrieking 'Geronimo!', like they did the last time, before the police came.

His solution is to throw a tasteful soiree. He has 26 different lagers, and he's only going to offer a six-course beer meal, where each beer will be accompanied by specially chosen potato chips and assorted snacks. How many possible drinking menus can Prokop offer, that is, how many different ways of picking out 6 beers from 20 are there?

The answer is quite simple. There are 26 different ways to start, all the way from Pilsen Amber to Pilsen Zoo. Then there are 25 ways to continue after that, 24 after that, 23, 22, 21. Now we have chosen six beers. So the answer is $26 \times 25 \times 24 \times 23 \times 22 \times 21$.

More generally, we start with n objects, continue with $n - 1$, $n - 2$, etc., but stop at $n - x + 1$. Conventionally this is written as

$$_nP_r = n(n-1)(n-2)\ldots(n-r+1)$$

where $_nP_r$ can be read as 'given n permute r'. At least, that's how I read it. This way of choosing is called permuting: $_nP_r$ computes the amount of permutations. An equivalent, but easier to use and remember, form is

$$_nP_r = \frac{n!}{(n-r)!}$$

obtained by multiplying our original formula by $(n-r)!/(n-r)!$

But Prokop comes to his senses—that's too damn much work. People can drink the beer in any order they wish. Perhaps that will stop the cursing about his habanero-stuffed jalapeños, which everyone but Matilda says are 'too hot'.

A.4.2 Combinations

This smoothly leads us to how to combine things when order is not important. The answer is fairly easy. We want to reduce the number of permutations: all sequences with the same number occurrences of objects count. But then we know that there are for any r objects $r!$ different ways to order them. We only want one way. So, we divided the number of permutations by $r!$, giving us

$$_nC_r = \frac{n!}{r!(n-r)!}.$$

(I read this as 'n choose r' but I don't know if anyone else does.) You might have noticed by now that 'combinatorics' is the study of combinations. It can be quite useful.

A.5 Laws of Large Numbers

This appendix is dedicated to the mathematical description of repetitions of independent trials. This requires first a way to represent the trials, which requires the combinatorics of the preceding section and the random variables of A.3.

A.5.1 Bernoulli random variables and the binomial distribution

The Bernoulli random variable takes just two values, 0 and 1. $p(X = 1) = 1 - p(X = 0)$. This is obviously perfect for modelling coin tosses. (The Bernoulli random variable is, obviously, the simplest non-trivial random variable, trivial random variables taking a constant value.)

Suppose that we will, as usual, toss a coin for some time, say n times, and we would like to know the probability that the coins land heads, say, r times, where the coin tosses are independent and with constant probability. Following tradition we call the occurrence of the phenomenon of interest a *success*, and refer to r successes.

Let us proceed in stages. First, we can determine the probability of a *particular* sequence coming up heads r times. Let us denote, in a slight abuse of notation, the probability of heads as p. Let's look at the sequence H, T, H, T, T ... What's the probability of this? It's $p \times (1-p) \times p \times (1-p) \times (1-p)$ ... Since the tosses are independent, the order doesn't matter. We can therefore rearrange the sequence as HHTTT. Since the tosses are independent, the probabilities multiply: $p \times p \times (1-p) \times (1-p) \times (1-p)$, or $p^2(1-p)^3$.

This can be easily generalized: the chance of any particular string coming up with r successes out of n is $p^r(1-p)^{r-n}$. But, we're not interested in any particular string, but of all strings with r successes. The number of strings with r successes out of n is given just a few paragraphs earlier. Putting these two together we get

$$p(X=r)=\frac{n!}{r!(n-r)!}p^r(1-p)^{n-r}$$

(The usual practice is to leave it implicit on the left-hand side that there are n trials.)

A practical, and important, example can be found in Prokop's extensive beer cellar. He allows Jarda to choose six beers at random (the light in the cellar is broken, it's really quite dark, and Prokop, not being in the least anally retentive, has no organizational system). Jarda, cursing messy people, is thirsty, and wants a lager—so picking a lager will now serve as a successful trial. Prokop and Jarda together have determined that the chance of an individual beer being a lager is $p = .4$. (How? Read the book to find out.) What is the chance of Jarda picking out exactly one lager, given that he picks six beers? We could take a complicated route. The formula gives us the chance that Jarda picks exactly one lager:

$$\begin{aligned} p(1 \text{ lager from } 6 \text{ beers}) &= \frac{6!}{1!(6-1)!}.4^1(1-.4)^{6-1} \\ &= \frac{6!}{5!}.4\times.07776=.186624 \end{aligned}$$

We can use this to determine the probability that Jarda will pick out *at least one* lager, that is, the chance that he draws one or two or three or four or five or six lagers. This is, using the obvious notion,

$$p(L=1)+p(L=2)+p(L=3)+p(L=4)+p(L=5)+p(L=6).$$

A faster route is simply:

$$1-p(L=0)=.953344$$

But what's the chance of Jarda's really quenching his thirst with at least three lagers? This is

$$p(L\geq 3)=p(L=3)+p(L=4)+p(L=5)+p(L=6)$$

Having done the calculations (or, rather, having used Excel), the values for $p(L=x)$ are

$p(L = 0) = .046656$

$p(L = 1) = .186624$

$p(L = 2) = .31104$

$p(L = 3) = .27648$

$p(L = 4) = .13824$

$p(L = 5) = .036864$

$p(L = 6) = .004096$

And so we can calculate that the probability of drawing three lagers, and maybe more, is .45568.

It might be instructive to contemplate the shape of the distribution of the values (Figure A.1), that is, of the probability distribution function:

The 'skew' of the distribution is due to the probability of failure being larger than that of success.

I could have been boring and given you an example with .5. Why boring? Figure A.2 shows how the distribution would be symmetrical around the mean value of the possibilities.

There are a number of (free!) programmes for visualizing binomial distributions with varying values of p, n, and r.

A.5.2 Laws of Large Numbers

Laws of Large Numbers describe, as you might expect, the (probabilistic) behaviour of outcomes of experiments as the number of experiments grows very large. They are provable from the axioms, showing how theorems of

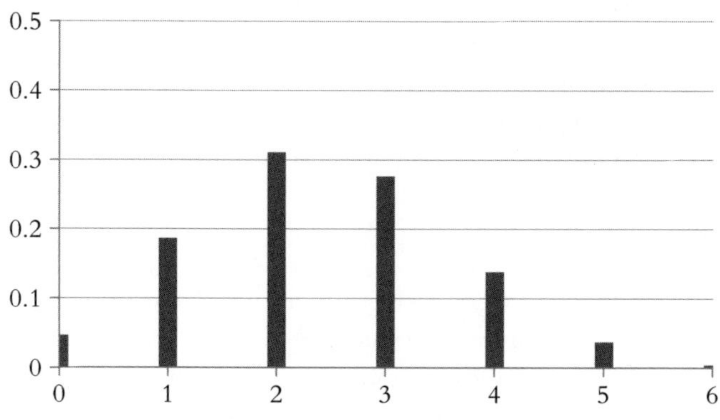

Figure A.1 Binomial distribution of successes in six trials, $p = .4$

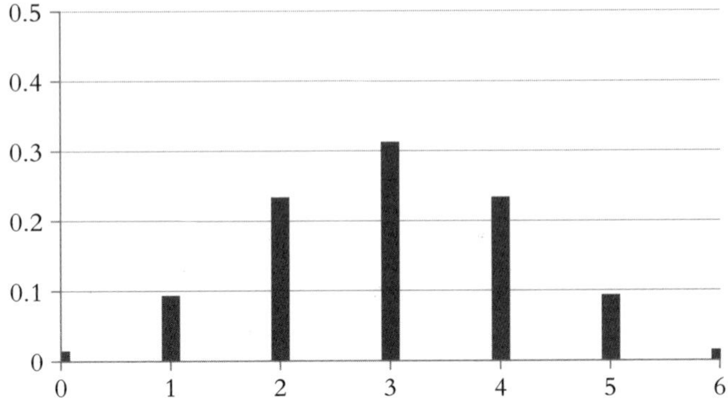

Figure A.2 Binomial distribution of successes in six trials, $p = .5$

great power can grow from simple foundations. The simplest such Laws involve experiments undertaken in such a way that they don't 'interfere' with each other, and are repeated exactly. The example is, of course, coin tossing. How does the relative frequency of the coin tosses $\frac{S_n}{n}$ behave as n gets (arbitrarily) large?

One answer is the *Weak Law of Large Numbers*, which tells us that the relative frequency of successes will settle down around the mean, the expected value μ, with probability one:

$$\lim_{n\to\infty} p\left(\left|\frac{S_n}{n} - \mu\right| < \varepsilon\right) \to 1,$$

for all ε. (This kind of convergence is known as convergence in probability, and so the Weak Law can be restated as saying that the relative frequency converges in probability to the mean.)

Another answer is the *Strong Law of Large Numbers*, so called because it is stronger. The Strong Law shows that not only does the relative frequency settle down around the mean—it will equal the mean:

$$p\left(\lim_{n\to\infty} \frac{S_n}{n} = \mu\right) = 1.$$

This kind of convergence is termed 'almost sure', and the Law can be restated as 'The relative frequency converges to the mean almost surely'. The Strong Law is significantly more difficult to prove than the Weak, and is perhaps of more philosophical interest, since it necessarily involves strong notions of infinity.

There are other theorems related to the Strong Law that provide estimates of the speed of convergence of relative frequencies to the mean. One of these theorems is the Law of the Iterated Logarithm. We will not even state this law, but refer the interested reader to the still-unmatched Feller 1957, chapter VIII, section 5. The Law states how often deviations from the mean of a certain size can occur, namely, if they're too big, only finitely, and if they're small enough, infinitely. And it gives precise meaning to 'too big' and 'too small'. (The Strong Law can also be read as being about the allowable amount of fluctuations from the mean in this sense, while the Weak Law cannot.)

A.5.3 Behaviour of the binomial distribution for large numbers of trials

Figure A.3 shows how the binomial distribution behaves for a fairly small amount of trials with probability .5 and number of trials equal to 10, 20, and 30. As can be seen, the probability starts to cluster around the mean. At first there are only 10 possible combinations of success, but the probabilities are quite spread out:

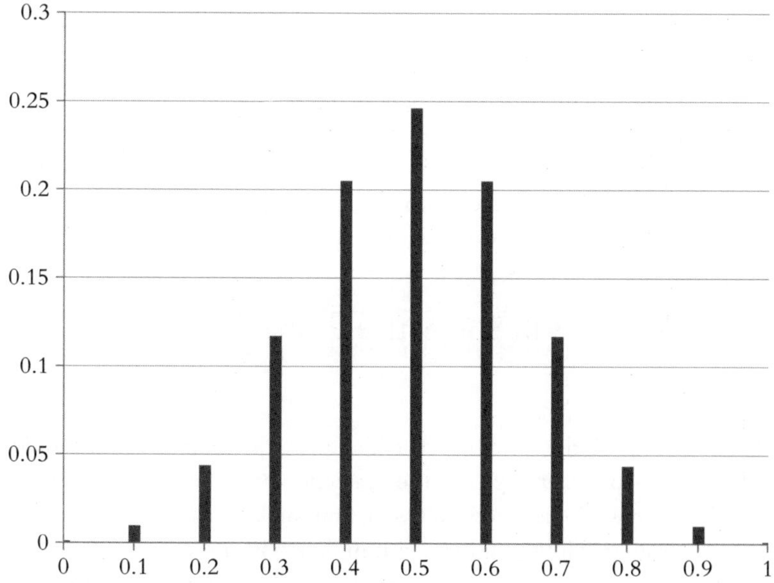

Figure A.3 Binomial distribution, percentage of success in 10 trials, $p = .5$

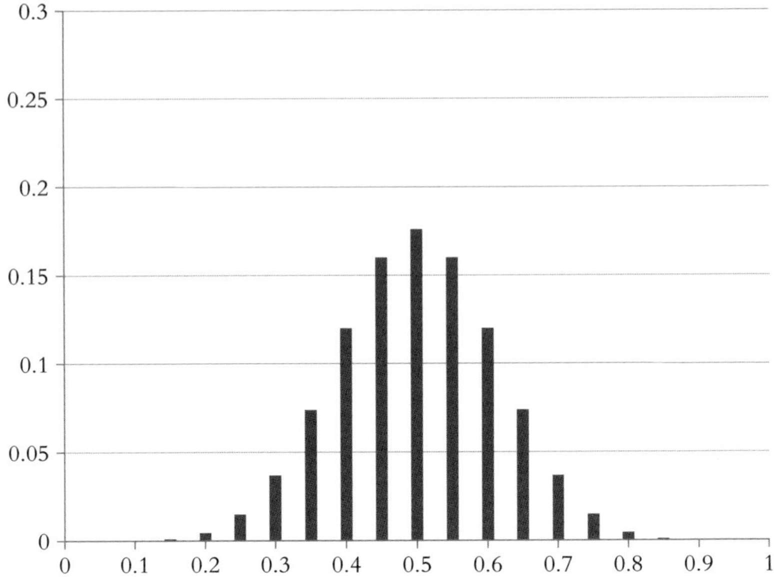

Figure A.4 Binomial distribution, percentage of success in 20 trials, $p = .5$

As Figure A.4 shows, after 20 trials the edges have dropped dramatically in probability.

After 30 they've more or less disappeared (Figure A.5).

As the number of trials increases, the probability of any particular percentage of success becomes very small. However, the probability of success remains clustered around the mean. (Don't let appearances fool you: the probability sums to one in all of the cases charted.)

A.6 Topics in Subjective Probability

A.6.1 Strict coherence

Strict coherence is the condition that fair betting quotients of 1 or 0 should only be assigned to tautologies and contradictions (in other words, never bet 1 or 0 if there is a logical possibility that you might lose). The justification seems appealing: why should you risk money with no chance of a gain? But this is what you do if you assign probability 1 to a non-tautologous (or non-necessary) statement.

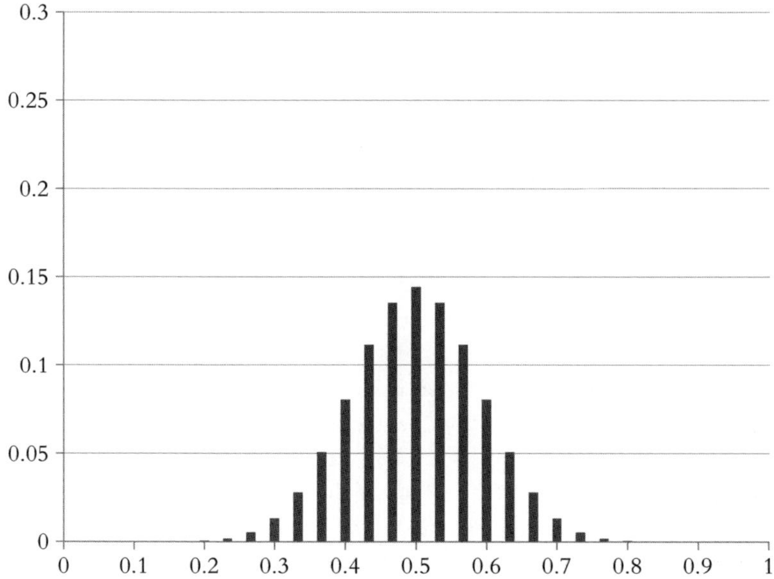

Figure A.5 Binomial distribution, percentage of success in 30 trials, $p = .5$

Making this criteria work, however, is somewhat more difficult. In fact, it can only be consistently applied if you have a finite amount of beliefs or if your probability measure is not countably additive. For if it is countably additive, and you want to assign a positive probability to each non-tautological and non-contradictory proposition, your probability will sum to more than one. (See A.2.4.1.)

A.6.2 Scoring rules

Prokop makes the acquaintance of the most fetching Matilda, a graduate student in finance. She works part-time at a major bank as a forecaster, although of what she won't tell him. On a walk through the park one day she mentions that her bosses tried to motivate her to give accurate forecasts about the possibility of an 'Oh my god we're all going to die' event in the next year. (Matilda shouted 'Oh my god we're all going to die', causing heads to turn. Prokop finds this delightful, but holds back saying so knowing that she would make a gagging noise in response. Instead he concentrates on the 'Mom' tattoo on her left bicep and blushes a Frankovka red.) If the event happens, she would get x dollars as a bonus, and if not, she would get 1,000 – x dollars. It's easy to see that $x/1,000$ is a kind of subjective probability (or can be made into one if we require it to add across events in the proper way). If her belief that it will happen is very strong, she will set x close to 1,000, and if her belief that it won't happen is very strong, she will set x close to 0. Or so her bosses think.

She pointed out to her bosses that even though she is a graduate student in finance she is highly risk averse, and downright lazy to boot. If she gave $x = 500$, she would be guaranteed 500 no matter what happened, and not have to work at all. But, she continued, since she is generally a nice person, and in an extra good mood, she would try hard to give a reasonable value to x. She laughs loudly, setting her piercings a-tinkle. Prokop's heart skips a beat.

Matilda's bosses employed what is known as a scoring rule: a device to elicit probability statements from people and at the same time to motivate them to give accurate forecasts. They attempted to take the difference between the occurrence of the event and the forecast of the event: this gives a number between 0 and 1, where 0 is the best score and 1 the worst. The reward is then the score times the 1,000 dollars. Let's take S for the reward, and L for the scoring rule. In this case $L = |E - p|$, where p is the normalized forecast, E is the indicator variable of the event, i.e. it is 1 if the event happens, 0 otherwise. Putting it together, we get that if A happened she would get $S|E - p|$. This gives us a suggestive pay-off table:

A	
T	$S(1-p)$
F	Sp

From an economist's point of view, if p were greater than .5, i.e. Matilda forecasts that the event would happen, and the event did *not* happen, then Sp is in fact a loss (an opportunity cost), since she could have got the larger sum $S(1-p)$ by forecasting more accurately. (The scoring rule for multiple events is divided by the number of events.)

But since we are dealing with Matilda, who is rational and well educated, she views her free time as worth something as well: for her there is no loss. So she is motivated to set $p = .5$ no matter what her personal views of the matter in fact are. This scoring rule is *improper*, since it does not motivate forecasters to reveal their 'true' probabilities, at least from the point of view of utility theory. If the forecaster is motivated to maximize his utility, and, unusually, equates the utility and value of money (for him the value of money is linear), then he will be motivated to pick either 1 or 0 depending on whether or not he thinks *A* will happen. (This is given as problem 6.11.8 in French 1988: 257, which is thankfully worked out on pages 409–10. This discussion is indebted to his account of scoring rules.)

Yet we don't need the machinery of utility theory to work out that this scoring rule doesn't encourage us to reveal our true probabilities. The following table gives values for pay-offs at various values of *p* when *A* does or doesn't happen.

p	*A*	Not-*A*
1	1000	0
.9	900	100
.8	800	200
.7	700	300
.6	600	400
.5	500	500
.4	400	600
.3	300	700
.2	200	800
.1	100	900
0	0	1000

Depending, of course, on the proposition, I feel a strong pull to the centre, .5.

One way to fix this flaw is to use another measure of distance. Three are current in the literature: the quadratic scoring rule, the log, and the spherical. We will only discuss the quadratic scoring rule since what follows also applies to the other two. The quadratic score is the square of the linear score $(E - p)^2$, and is proper: maximizing your expected utility, no matter what shape the curve describing it is, requires stating your true probabilities. (This is also worked out in French, in the next problem 6.11.9.) Still, if we look at the table of pay-offs,

P	*A*	Not-*A*
1	1000	0
.9	990	190
.8	960	360
.7	910	510
.6	840	640
.5	750	750
.4	640	840
.3	510	910
.2	360	960
.1	190	990
0	0	1000

we can see that our previous problem recurs in an even worse form: laziness is well rewarded, hence the quadratic scoring rule need not reveal the subjective probabilities of a risk-averse or lazy subject.

A.6.3 Axioms for qualitative and quantitative probabilities

The following is drawn from French 1988 and 1982, who follows DeGroot 1970. (Childers and Majer 1999 is also used to provide a slight modification.) As usual, we assume that there is an algebra F of events over Ω with standard set theoretic operations. Then we define a relation of relative likeliness over real-world events, $A \gtrsim B$, meaning that A is at least as likely as B. The first three axioms are those of qualitative probability:

Weak Ordering

$\gtrsim$ is a weak order over the elements of F.

That is for all $A, B, C \in F$, (i) $A \gtrsim B$ and/or $B \gtrsim A$, (ii) if $A \gtrsim B$ and $B \gtrsim C$ then $A \gtrsim C$. $>$ and $\sim$ can now be defined in the usual way, e.g. $A \sim B$ if and only if $A \gtrsim B$ and $B \gtrsim A$, $A > B$ if and only if it is not the case that $B \gtrsim A$.

Independence of Common Events

For any $A, B, C \in \Omega$, if $A \cap C = \varnothing = B \cap C$ then

$A \gtrsim B \Leftrightarrow A \cup C \gtrsim B \cup C$

Non-Triviality

$\Omega > \varnothing, A \gtrsim \varnothing$, for any $A \in \Omega$.

To calibrate the likeliness of our events we need a scaling device. While any fair game of fortune will do, we will stick with the wheel of fortune. We only require that we can build an algebra of events from the game, and define a flat distribution over the elementary events.

A wheel of fortune can be taken as a unit circle centred on the origin of a polar coordinate system. Any arc can be expressed as a pair $[\alpha, \beta]$, where α determines the start of the arc on the circle and β the second (or we could take β to determine the length of the arc). The set of all arcs of the circle determines an algebra of reference events.

Reference Experiment

The reference experiment consist of events of the algebra G generated by the set $\{[\alpha_0, \varphi]; \alpha_0, \varphi \in \langle 0, 2\pi \rangle\}$ with the usual set-theoretic operations.

Relative likelihood in the reference experiment is defined using length (since we've assumed it's fair). Denoting length as $l(A)$,

Relative Likelihood for the Reference Experiment

$$A \succsim_R B \Leftrightarrow l(A) \geq l(B),$$

$\succsim_R$ satisfies the axioms of qualitative probability (since length, being a measure, does).

Relating the events with their reference events requires a continuity assumption. This assumption is crucial, since it is how the probabilities will be determined.

Continuity

The likelihood orderings $\succsim$ over F and $\succsim_R$ over G can be extended to a likelihood ordering over the Cartesian product of F and G *and* for any A in F, there is a $A' \in G$ such that $A \sim A'$.

The last assumption determines a convention matching the scales.

Equivalence of Certainties

$$\Omega \sim [0, 2\pi]$$

The probability is now the reference event, the normalized length of the arc, corresponding to the real-world event. It can be proved (French 1988) that

$p(A) = f(A)/2\pi$ is a probability measure over the field F.

This completes the construction: the wheel of fortune gives probabilities that completely determine probabilities over the primitive likeliness ordering.

A.7 The Duhem-Quine Problem, Language, Metaphysics

W.V. Quine produced a remarkably comprehensive empiricist philosophy in which central intertwining themes like holism (yielded by a Duhem-Quine argument applied to an empiricist theory of language) and ontological relativity provided a working out of a thoroughgoing naturalism. The key to understanding this view is to grasp Quine's austere version of language learning—for him, empiricism is encapsulated in the view that we can only depend on what others around us signal if we want to understand a language.

This is not outlandish. We can imagine a child learning this way, and we can certainly picture ourselves in the position of someone dropped in the middle of a rural unconnected backwater trying to learn a language. A dream: Prokop has been kidnapped and dumped in the middle of the Atchafalaya basin, where he is taken in by Gaston Boudreaux, a kind-hearted Cajun. One early morning as he helps his host Gaston rig catfish lines across the bayou, a rabbit hops into view on the opposite bank. Gaston points and whispers 'lapin'. Prokop can't understand Gaston's heavily accented English most of the time. Nor does Prokop speak French, although he doubts it would help much in understanding Gaston's dialect. But he suspects that Gaston is saying 'rabbit'. If his hands weren't covered with foul-smelling catfish bait he would get out his battered moleskin and write 'rabbit = lapin'.

But would he be justified in doing so? Gaston and his fellow Cajuns might be so taken with cooking rabbits in various ways (and from my authorial perspective I assure you that they are) that he could actually be saying 'undetached rabbit parts' or 'very tasty rabbit meat once cleaned'. Perhaps they have adopted a Minkowskian ontological framework, and Gaston is actually saying 'time-slice of a rabbit'. You might argue that Prokop could, by following Gaston around enough, manage to rule out these seemingly outlandish interpretations. Not so, argues Quine. If we take the only evidence for the meaning of words to be the empirical context in which they are uttered, we will find that the Duhem-Quine problem always leaves us with ways to fiddle with our translations. They might be outlandish—portraying Cajuns wholeheartedly adopting a view of medium-sized objects inspired by Minkowskian space-time as opposed to wholeheartedly adopting one inspired by various ways of preparing objects for consumption—but they won't be dismissible on the grounds that they lack evidence. This example serves to show that, given this austerely empirical theory of meaning (which, by the way, is due to B.F. Skinner), there can never be a matter of fact as to how a language is related to its semantics, where semantics is taken to imply a particular ontology.

Finally Gaston puts Prokop in his pirogue, poles him to the nearest town with a bus station, buys him a ticket to California, kisses him on both cheeks, and sends him off with a Tupperware container of lapin étouffée.

A.7.1 A probabilistic translation of Quine's programme

The probabilistic translation is as follows: meanings of words are determined by conditional probabilities, $p(\textit{utterance} \mid \textit{stimulus}, \textit{environment})$. So, if the probability *p*(Gaston says 'lapin' | rabbit in view) is high, then 'lapin' has the meaning 'rabbit'. (A caution: I play fast and loose with 'meaning'. Quine uses the term 'stimulus meaning', and this, it turns out, is not a traditional notion of meaning.) But this formulation leaves something out. For the implication really should be *p*(Gaston says 'lapin' | rabbit in view & Gaston shares Prokop's metaphysics) is high, therefore 'lapin' means 'rabbit'. But the conditional probability is (or can be) equal to *p*(Gaston says 'lapin' | rabbit in view & Gaston has a food-obsessed metaphysics), and from this we can conclude 'lapin' means 'undetached rabbit parts'. According to Quine, if we are willing to be flexible, no empirical evidence can differentiate between the two. (Note: this is not a problem of empirical underdetermination. It holds even if we had all available empirical evidence.) Since, for Quine, semantics is nothing other than confirmation, the conclusion is inescapable: there is no fact of the matter as to the meaning of the sentence, and hence to the metaphysics of Gaston's world view. (Some claim that we should distinguish between semantic and conformational holism. For Quine there is no such distinction.)

Quine reworks intertwined themes in his own, very unique, style of American English throughout his corpus. An accessible yet reliable guide to his thought is Kemp 2006. The themes I have been surveying are explicitly worked out in Quine's 1960 *Word and Object*. His 'Ontological Relativity', the second article in his 1969 collection of the same name, is also a useful guide. Quine's 1981 'Reply to Roth' is useful as a map.

References

Albert, Max. 2005. Should Bayesians Bet where Frequentists Fear to Tread? *Philosophy of Science* 72: 584–93.

Anand, Paul. 1993. *Foundations of Rational Choice Under Risk*. Oxford: Oxford University Press.

Armendt, Brad. 1993. Dutch Books, Additivity and Utility Theory. *Philosophical Topics* 21(1): 1–20.

Ash, Robert B. 1965. *Information Theory*. New York: Dover Publications, Inc.

Bass, Thomas A. 1991. *The Newtonian Casino*. London: Penguin.

Bernoulli, Jacob. (1713) 2006. *The Art of Conjecturing: Together with 'Letter to a Friend on Sets in Court Tennis'*. Translated by Edith Dudley Sylla. Baltimore: Johns Hopkins University Press.

Billingsley, Patrick. 1995. *Probability and Measure*. 3rd edn. New York: John Wiley & Sons.

Binmore, Ken. 2009. *Rational Decisions*. Princeton: Princeton University Press.

Borel, Émile. (1950) 1965. *Elements of the Theory of Probability*. Translated by John E. Freund. Englewood Cliffs: Prentice-Hall.

Bovens, Luc and Stephan Hartmann. 2003. *Bayesian Epistemology*. Oxford: Oxford University Press.

Briggs, Rachael. 2009a. The Big Bad Bug Bites Anti-realists about Chance. *Synthese* 167: 81–92.

——. 2009b. The Anatomy of the Big Bad Bug. *Noûs* 43(3): 428–49.

Brukner, Časlav and Anton Zeilinger. 2001. Conceptual Inadequacy of the Shannon Information in Quantum Measurements. *Physical Review A* 63(2): 022113.

Callender, Craig and Jonathan Cohen. 2010. Special Sciences, Conspiracy and the Better Best System Account of Lawhood. *Erkenntnis* 73: 427–47.

Carnap, Rudolf. (1932) 1959. The Elimination of Metaphysics through the Logical Analysis of Language. In *Logical Positivism*, 60–81. Translated by Arthur Pap. Edited by A. Ayer. New York: The Free Press.

——. (1934) 1967. On the Character of Philosophic Problems. In *The Linguistic Turn: Recent Essays in Philosophical Method*, 54–62. Translated by W.M. Malisof. Edited by Richard Rorty. Chicago: University of Chicago Press.

——. 1950. *Logical Foundations of Probability*. Chicago: University of Chicago Press.

——. 1952. *The Continuum of Inductive Methods*. Chicago: University of Chicago Press.

Childers, Timothy. 2009. After Dutch Books. In *Foundations of the Formal Sciences VI: Reasoning about Probabilities and Probabilistic Reasoning*, 103–15. Edited by B. Löwe, E. Pacuit, and J.-W. Romeijn. London: College Publications London.

——. 2012. Dutch Book Arguments for Direct Probabilities. In *Probabilities, Laws, and Structures: The Philosophy of Science in a European Perspective* Vol. 3, 19–28. Edited by D. Dieks, W. Gonzales, S. Hartmann, M. Stöltzner, and M. Weber. Berlin: Springer.

—— and Ondrej Majer. 1998. Łukasiewicz's Theory of Probability. In *The Lvov-Warsaw School and Contemporary Philosophy*, 303–12. Edited by K. Kijania-Placek and J. Woleński. Dordrecht: Kluwer.

—— ——. 1999. Representing Diachronic Probabilities. In *The Logica Yearbook 1998*, 170–9. Edited by T. Childers. Prague: Filosofia.

Christensen, David. 1991. Clever Bookies and Coherent Beliefs. *The Philosophical Review* 100(2): 229–47.

——. 1999. Measuring Confirmation. *The Journal of Philosophy* 96(9): 437–61.

Church, Alonzo. 1940. On the Concept of a Random Sequence. *Bulletin of the American Mathematical Society* 46: 130–5.

Colyvan, Mark. 2004. The Philosophical Significance of Cox's Theorem. *International Journal of Approximate Reasoning* 37: 71–85.

Connors, Edward, Thomas Lundregan, Neal Miller, and Tom McEwen. 1996. *Convicted by Juries, Exonerated by Science: Case Studies in the Use of DNA Evidence to Establish Innocence After Trial.* Research Report NCJ 177626. Washington, DC: National Institute of Justice.

Cover, Thomas M. and Joy A. Thomas. 2006. *Elements of Information Theory.* 2nd edn. New York: John Wiley & Sons.

Cox, Richard T. 1946. Probability, Frequency and Reasonable Expectation. *American Journal of Physics* 14(1): 1–13.

——. 1961. *The Algebra of Probable Inference.* Baltimore: Johns Hopkins University Press.

Deakin, Michael. 2006. The Wine/Water Paradox: Background, Provenance and Proposed Resolutions. *The Australian Mathematical Society Gazette* 33: 200–5.

de Finetti, Bruno. (1931) 1993. On the Subjective Meaning of Probability. In *Probabilità e Induzione*, 291–321. Edited by P. Monari and D. Cocchi. Bologna: Clueb.

——. (1937) 1964. Foresight: Its Logical Laws, Its Subjective Sources. In *Studies in Subjective Probability*, 97–158. Translated by Henry Kyburg. Edited by Henry Kyburg and Howard Smokler. New York: John Wiley & Sons.

——. 1972. *Probability, Induction and Statistics: The Art of Guessing.* London: John Wiley & Sons.

——. 1974. *Theory of Probability: A Critical Introductory Treatment*, Vol. 1. Translated by Antonio Machí and Adrian Smith. London: John Wiley & Sons.

DeGroot, Morris H. 1970. *Optimal Statistical Decisions.* New York: McGraw-Hill.

Diaconis, Persi and David Freedman. 1986. On the Consistency of Bayes Estimates. *Annals of Statistics* 14: 1–26.

—— Susan Holmes, and Richard Montgomery. 2007. Dynamical Bias in the Coin Toss. *SIAM Review* 49(2): 211–35.

Doob, J.L. 1941. Probability as Measure. *The Annals of Mathematical Statistics* 12(3): 206–14.

Dorling, Jon. 1979. Bayesian Personalism, the Methodology of Scientific Research Programmes, and Duhem's Problem. *Studies in History and Philosophy of Science* 10(3): 177–87.

Downey, Rodney G. and Denis R. Hirschfeldt. 2010. *Algorithmic Randomness and Complexity*. Heidelberg: Springer.

Eagle, Anthony. 2004. Twenty-one Arguments against Propensity Analyses of Probability. *Erkenntnis* 60: 371–416.

——. 2012. Chance versus Randomness. In *The Stanford Encyclopedia of Philosophy*, Spring 2012 Edition. Edited by Edward N. Zalta. <http://plato.stanford.edu/archives/spr2012/entries/chance-randomness/>.

Earman, John. 1992. *Bayes or Bust?: A Critical Examination of Bayesian Confirmation Theory.* Cambridge, MA: MIT Press.

Economic Commission for Europe. 2007. *Statistics of Road Traffic Accidents in Europe and North America*, Vol. 51. New York: United Nations. <http://www.unece.org/trans/main/wp6/pdfdocs/RAS_2007.pdf>.

Edwards, Ward, Harold Lindman, and Leonard J. Savage. 1963. Bayesian Statistical Inference for Psychological Research. *Psychological Review* 70(3): 193–242.

Eells, Ellery and Branden Fitelson. 2000. Measuring Confirmation and Evidence. *The Journal of Philosophy* 97(12): 663–72.

Feller, William. 1957. *An Introduction to Probability Theory and Its Applications*. 2nd edn. New York: John Wiley & Sons.

Fetzer, James. 1981. *Scientific Knowledge: Causation, Explanation, and Corroboration.* Dordrecht: D. Reidel Publishing Co.

Fienberg, Stephen E. 2006. When Did Bayesian Inference Become 'Bayesian'? *Bayesian Analysis* 1(1): 1–40.

Fishburn, Peter. 1986. The Axioms of Subjective Probability. *Statistical Science* 1(3): 335–45.

Fitelson, Branden. 2001. *Studies in Bayesian Confirmation Theory*. PhD dissertation, University of Wisconsin-Madison. <http://www.fitelson.org/thesis.pdf>.

—— and Andrew Waterman. 2005. Bayesian Confirmation and Auxiliary Hypotheses Revisited: A Reply to Strevens. *British Journal for the Philosophy of Science* 56(2): 293–302.

Föllmer, Hans and Uwe Küchler. 1991. Richard von Mises. In *Mathematics in Berlin*, 111–16. Edited by H.G.W. Begehr, H. Koch, J. Kramer, N. Schappacher, E.-J. Thiele. Berlin: Birkhäuser Verlag.

Franklin, James. 2001. *The Science of Conjecture: Evidence and Probability before Pascal.* Baltimore: Johns Hopkins University Press.

Fréchet, Maurice. 1939. The Diverse Definitions of Probability. *Erkenntnis* 8(1): 7–23.

French, Simon. 1982. On the Axiomatisation of Subjective Probabilities. *Theory and Decision* 14: 19–33.

——. 1988. *Decision Theory: An Introduction to the Mathematics of Rationality.* Chichester: Ellis Horwood.

Galavotti, Maria Carla. 2005. *Philosophical Introduction to Probability.* Stanford: CSLI Publications.

Geiringer, Hilda. 1969. Probability Theory of Verifiable Events. *Archive for Rational Mechanics and Analysis* 34(1): 3–69.

Giere, Ronald N. 1976. A Laplacean Formal Semantics for Single-Case Propensities. *Journal of Philosophical Logic* 5: 321–53.

Gillies, Donald A. 1973. *An Objective Theory of Probability.* London: Methuen & Co. Ltd.

——. 1990. Bayesianism Versus Falsificationism. *Ratio* 3(1): 82–98.

——. 2000. *Philosophical Theories of Probability.* London: Routledge.

Guay, Alexandre and Brian Hepburn. 2009. Symmetry and Its Formalisms: Mathematical Aspects. *Philosophy of Science* 76(2): 160–78.

Hacking, Ian. 2001. *An Introduction to Probability and Logic.* Cambridge: Cambridge University Press.

Hájek, Alan. 1997. 'Mises Redux' – Redux: Fifteen Arguments against Finite Frequentism. *Erkenntnis* 45(2, 3): 209–27.

——. 2003. What Conditional Probability Could Not Be. *Synthese* 137(3): 273–323.

——. 2007. The Reference Class Problem Is Your Problem Too. *Synthese* 156(3): 563–85.

——. 2009. Fifteen Arguments against Hypothetical Frequentism. *Erkenntnis* 70: 211–35.

——. 2012. Interpretations of Probability. *The Stanford Encyclopedia of Philosophy*, Summer 2012 Edition. Edited by Edward N. Zalta. <http://plato.stanford.edu/archives/sum2012/entries/probability-interpret/>.

Hall, Michael J.W. 2000. Comment on 'Conceptual Inadequacy of Shannon Information ...' by C. Brukner and A. Zeilinger. <http://arxiv.org/abs/quant-ph/0007116v1>.

Hall, Ned. 1994. Correcting The Guide to Objective Chance. *Mind* 103(412): 505–17.

Hayles, N. Katherine. 1999. *How We Became Posthuman: Virtual Bodies in Cybernetics, Literature, and Informatics.* Chicago: University of Chicago Press.

Heath, David and William Sudderth. 1976. De Finetti's Theorem on Exchangeable Variables. *The American Statistician* 30(4): 188–9.

Hoefer, Carl. 2007. The Third Way on Objective Probability: A Sceptic's Guide to Objective Chance. *Mind* 116(463): 549–96.

Howson, Colin. 1995. Theories of Probability. *The British Journal for the Philosophy of Science* 46(1): 1–32.

——. 2000. *Hume's Problem: Induction and the Justification of Belief*. Oxford: Oxford University Press.

——. 2003. Probability and Logic. *Journal of Applied Logic* 1: 151–65.

——. 2008. De Finetti, Countable Additivity, Consistency and Coherence. *British Journal for the Philosophy of Science* 59: 1–23.

——. 2009. Can Logic Be Combined with Probability? Probably. *Journal of Applied Logic* 7(2): 177–87.

—— and Peter Urbach. 1993. *Scientific Reasoning: The Bayesian Approach*. 2nd edn. Chicago: Open Court.

—— ——. 2006. *Scientific Reasoning: The Bayesian Approach*. 3rd edn. Chicago: Open Court.

Humphreys, Paul. 1982. Review of *Hans Reichenbach: Logical Empiricist*. *Philosophy of Science* 49(1): 140–2.

——. 1985. Why Propensities Cannot Be Probabilities. *The Philosophical Review* 94(4): 557–70.

——. 2004. Some Considerations on Conditional Chances. *British Journal for the Philosophy of Science* 55: 667–80.

Jaynes, E.T. 1957a. Information Theory and Statistical Mechanics. *The Physical Review* 106(4): 620–30.

——. 1957b. Information Theory and Statistical Mechanics II. *The Physical Review* 108(2): 171–90.

——. 1968. Prior Probabilities. *IEEE Transactions On Systems Science and Cybernetics* 4(3): 227–41.

——. 1973. The Well-posed Problem. *Foundations of Physics* 3: 477–93.

Jeffrey, Richard. 1983. *The Logic of Decision*. 2nd edn. Chicago: University of Chicago Press.

——. 1993. Take Back the Day! Jon Dorling's Bayesian Solution of the Duhem Problem. *Philosophical Issues* 3: 197–207.

Joyce, James. 1998. A Non-pragmatic Vindication of Probabilism. *Philosophy of Science* 65(4): 575–603.

Kemp, Gary. 2006. *Quine: A Guide for the Perplexed*. London: Continuum.

Keynes, John Maynard. 1921 (1973). *A Treatise on Probability*. Vol. 8 of *The Collected Writings of John Maynard Keynes*. London: Macmillan for the Royal Economic Society.

Kolmogorov, A.N. (1933) 1950. *Foundations of the Theory of Probability*. Translation edited by Nathan Morrison. New York: Chelsea Publishing Co.

——. 1963. On Tables of Random Numbers. *Sankhyā: The Indian Journal of Statistics, Series A* 25.

Kosso, Peter. 2000. The Empirical Status of Symmetries in Physics. *British Journal for the Philosophy of Science* 51(1): 81–98.

Kposowa, Augustine J. and Michele Adams. 1998. Motor Vehicle Crash Fatalities: The Effects of Race and Marital Status. *Applied Behavioral Science Review* 6(1): 69–91.

Kraft, Charles H., John W. Pratt, and A. Seidenberg. 1959. Intuitive Probability on Finite Sets. *Annals of Mathematical Statistics* 30: 408–19.

Kullback, S., and R.A. Leibler. 1951. On Information and Sufficiency. *The Annals of Mathematical Statistics* 22(1): 79–86.

Kyburg, Henry E. 1981. Principle Investigation. *Journal of Philosophy* 78(12): 772–8.

La Pensée, Clive. 1990. *The Historical Companion to House-Brewing*. Beverley: Montag Publications.

Laplace, Pierre-Simon, Marquis de. (1814) 1952. *Essai Philosophique sur les Probabilités*. Translated as *Philosophical Essay on Probabilities* by E.T. Bell, 1902. Reprint, New York: Dover Publications, Inc.

Laraudogoitia, Jon Pérez. 2011. Supertasks. In *The Stanford Encyclopedia of Philosophy*, Spring 2011 Edition. Edited by Edward N. Zalta. <http://plato.stanford.edu/archives/spr2011/entries/spacetime-supertasks/>.

Lewis, David. (1973) 1986. Counterfactuals and Comparative Possibility. In Lewis 1986, 3–31.

——. (1980) 1986. A Subjectivist's Guide to Objective Chance. In Lewis 1986, 83–113.

——. 1986a. *Philosophical Papers*, Vol. 2. Oxford: Oxford University Press.

——. 1986b. Introduction. In Lewis 1986, ix–xvii.

——. 1994. Humean Supervenience Debugged. *Mind* 103(412): 473–90.

——. 1999. Why Conditionalize? In his *Papers in Metaphysics and Epistemology*, 403–7. Cambridge: Cambridge University Press.

Li, Ming and Paul Vitanyi. 1997. *An Introduction to Kolmogorov Complexity and Its Applications*. 2nd edn. Berlin: Springer.

Lindley, D.V. 1953. Review of *Stochastic Processes* by J.L. Doob. *Journal of the Royal Statistical Society. Series A* 116(4): 454–6.

Luce, R. Duncan and Howard Raiffa. (1957) 1989. *Games and Decisions: Introduction and Critical Survey*. Reprint, New York: Dover Publications, Inc.

Maher, Patrick. 2010. Explication of Inductive Probability. *Journal of Philosophical Logic* 39(6): 593–616.

Mana, Piero G. Luca. 2004. Consistency of the Shannon entropy in quantum experiments. *Physical Review A* 69. 062108.

Maor, Eli. 1994. *e: The Story of a Number*. Princeton: Princeton University Press.

Marinoff, Louis. 1994. A Resolution of Bertrand's Paradox. *Philosophy of Science* 61(1): 1–24.

Martin-Löf, P. 1969. The Literature on von Mises' Kollektivs Revisited. *Theoria* 35(1): 12–37.

Mayo, Deborah G. 1996. *Error and the Growth of Experimental Knowledge*. Chicago: University of Chicago Press.

McCurdy, Christopher S.I. 1996. Humphreys' Paradox and the Interpretation of Inverse Conditional Propensities. *Synthese* 108(1): 105–25.

Mellor, D.H. 1971. *The Matter of Chance*. Cambridge: Cambridge University Press.

——. 2005. *Probability: A Philosophical Introduction*. London: Routledge.

Metzger, Bruce M. and Bart D. Ehrman. 2005. *The Text of the New Testament: Its Transmission, Corruption, and Restoration*. 4th edn. Oxford: Oxford University Press.

Mikkelson, Jeffrey M. 2004. Dissolving the Wine/Water Paradox. *British Journal for the Philosophy of Science* 55: 137–45.

Miller, David. 1994. *Critical Rationalism: A Restatement and Defence*. Chicago: Open Court.

Milne, Peter. 1983. A Note on Scale Invariance. *The British Journal for the Philosophy of Science* 34(1): 49–55.

——. 1986. Can There Be a Realist Single-case Interpretation of Probability? *Erkenntnis* 25: 129–32.

——. 1991. Annabelle and the Bookmaker. *Australasian Journal of Philosophy* 69(1): 98–102.

——. 1996. log[P(h/eb)/P(h/b)] Is the One True Measure of Confirmation. *Philosophy of Science* 63(1): 21–6.

——. 1997. Bruno de Finetti and the Logic of Conditional Events. *British Journal for the Philosophy of Science* 48: 195–232.

——. 2003. Bayesianism v. Scientific Realism. *Analysis* 63(4): 281–8.

Neyman, Jerzy. 1952. *Lectures and Conferences on Mathematical Statistics and Probability*. 2nd edn. Washington: US Department of Agriculture.

NHTSA (National Highway Traffic Safety Administration). 2011. *Traffic Safety Facts 2009: A compilation of motor vehicle crash data from the FARS and the GES*, Early edition, DOT HS 811 402. Washington, DC: NHTSA. <http://www-nrd.nhtsa.dot.gov/pubs/811402ee.pdf>. Volume undated, date inferred from other government publications.

Niiniluoto, Ilkka. 2011. The Development of the Hintikka Program. In *Inductive Logic*, 311–56. Edited by Dov M. Gabbay, Stephan Hartmann, and John Woods. Vol. 10 of the *Handbook of the History and Philosophy of Logic*. North-Holland: Amsterdam. <http://stephanhartmann.org/HHL10_Niiniluoto.pdf>.

Nolan, Daniel. 2005. *David Lewis*. Chesham: Acumen Publishing Ltd.

Paris, J.B. 1994. *The Uncertain Reasoner's Companion: A Mathematical Perspective*. Cambridge: Cambridge University Press.

—— and A. Vencovská. 2001. Common Sense and Stochastic Independence. In *Foundations of Bayesianism*, 203–40. Edited by D. Corfield, and J. Williamson. Dordrecht: Kluwer.

—— ——. 2011. Symmetry's End? *Erkenntnis* 74: 53–67.

—— ——. 2012. Symmetry in Polyadic Inductive Logic. *Journal of Logic, Language and Information* 21: 189–216.

Pettigrew, Richard. 2012. Accuracy, Chance, and the Principal Principle. *Philosophical Review* 121 (2): 241–75.

Popper, Karl R. 1959. The Propensity Interpretation of Probability. *The British Journal for the Philosophy of Science* 10 (37): 25–42.

Quine, W.V.O. (1953) 1980. Two Dogmas of Empiricism. In *From a Logical Point of View*, 2nd edn., 20–46. Cambridge, MA: Harvard University Press.

——. 1960. *Word and Object*. Cambridge, MA: The Technology Press of the Massachusetts Institute of Technology.

——. 1969. *Ontological Relativity and Other Essays*. New York: Columbia University Press.

——. 1981. Reply to Paul A. Roth. In *Midwest Studies in Philosophy*, 459–61. Edited by P.A. French, T.E. Uehling, and A.K. Wettstein. Minneapolis: University of Minnesota Press.

——. 1990. *Quiddities: An Intermittently Philosophical Dictionary*. London: Penguin Books.

Ramsey, Frank Plumpton (1926) 1931. Truth and Probability. In *The Foundations of Mathematics and other Logical Essays*, 156–98. Edited by R.B. Braithwaite. London: Kegan, Paul, Trench, Trubner & Co.

——. (1927) 1978. Facts and Propositions. In *Foundations: Essays in Philosophy, Mathematics and Economics*, 40–57. Edited by D.H. Mellor. London: Routledge & Kegan Paul.

Redhead, Michael. 1980. A Bayesian Reconstruction of the Methodology of Scientific Research Programmes. *Studies in History and Philosophy of Science* 11: 341–7.

Reichenbach, Hans. 1949. *The Theory of Probability: An Inquiry into the Logical and Mathematical Foundations of the Calculus of Probability*. Translated by E.H. Hutten and M. Reichenbach. 2nd edn. Berkeley and Los Angeles: University of California Press.

Romeijn, Jan-Willem. 2005. Theory Change and Bayesian Statistical Inference. *Philosophy of Science* 72: 1174–86.

Rosenberg, Alexander and Daniel W. McShea. 2008. *Philosophy of Biology: A Contemporary Introduction*. London: Routledge.

Rosenkrantz, Roger D. 1977. *Inference, Method and Decision: Towards a Bayesian Philosophy of Science*. Dordrecht: D. Reidel Publishing Co.

Savage, Leonard J. 1954. *The Foundations of Statistics*. 2nd edn. New York: Dover Publications, Inc.

——. 1971. Elicitation of Personal Probabilities and Expectations. *Journal of the American Statistical Association* 66 (336): 783–801.

Schick, Frederic. 1986. Dutch Bookies and Money Pumps. *Journal of Philosophy* 83(2): 112–19.

Seidenfeld, Teddy. 1979. Why I Am Not an Objective Bayesian: Some Reflections Prompted by Rosenkrantz. *Theory and Decision* 11: 413–40.

Shafer, Glenn and Vladimir Vovk. 2001. *Probability and Finance: It's Only a Game!* New York: John Wiley & Sons.

Shannon, Claude. 1948. A Mathematical Theory of Communication. *Bell System Technical Journal* 27: 379–423, 623–56.

Shore, John E. and Rodney W. Johnson. 1980. Axiomatic Derivation of the Principle of Maximum Entropy and the Principle of Minimum Cross-Entropy. *IEEE Transactions on Information Theory* 26(1): 26–37.

Sklar, Lawrence. 1993. *Physics and Chance: Philosophical Issues in the Foundations of Statistical Mechanics*. Cambridge: Cambridge University Press.

Skyrms, Brian. 1986. *Choice and chance: an introduction to inductive logic.* 3rd edn. Belmont, CA: Wadsworth Publishing Co.

——. 2012. Review of *Rational Decisions* by Ken Binmore. *British Journal for the Philosophy of Science* 63(2): 449–53.

Smith, Cedric A.B. 1961. Consistency in Statistical Inference and Decision. *Journal of the Royal Statistical Society: Series B* 23(1): 1–37.

Sober, Elliott. 2000. *Philosophy of Biology*. 2nd edn. Boulder, CO: Westview Press.

——. 2008. *Evidence and Evolution: The Logic Behind the Science*. Cambridge: Cambridge University Press.

Strevens, Michael. 1999. Objective Probability as a Guide to the World. *Philosophical Studies* 95: 243–75.

——. 2001. The Bayesian Treatment of Auxiliary Hypotheses. *British Journal for the Philosophy of Science* 52: 515–38.

——. 2003. *Bigger Than Chaos: Understanding Complexity through Probability*. Cambridge, MA: Harvard University Press.

——. 2005a. The Bayesian Treatment of Auxiliary Hypotheses: Reply to Fitelson and Waterman. *British Journal for the Philosophy of Science* 56(4): 913–18.

——. 2005b. Probability and Chance. *Encyclopedia of Philosophy.* 2nd edn. Macmillan Reference USA. <http://www.strevens.org/research/simplexuality/Probability.pdf>.

Teller, Paul. 1973. Conditionalization and Observation. *Synthese* 26(2): 218–58.

Thau, Michael. 1994. Undermining and Admissibility Source. *Mind* 103(412): 491–503.

Timpson, Christopher G. 2003. The Applicability of Shannon Information in Quantum Mechanics and Zeilinger's Foundational Principle. *Philosophy of Science* 70: 1233–44.

Turing, A.M. 1937. On Computable Numbers, with an Application to the *Entscheidungsproblem*. *Proceedings of the London Mathematical Society Series 2* 42: 230–65.

Uffink, Jos. 1990. *Measures of Uncertainty and the Uncertainty Principle*. PhD dissertation, Utrecht University. <http://www.phys.uu.nl/igg/jos/>.

——. 1996. Can the Maximum Entropy Principle Be Explained As a Consistency Requirement? *Studies in History and Philosophy of Modern Physics* 26(3): 223–61.

van Fraassen, Bas. 1980. *The Scientific Image*. Oxford: Oxford University Press.

——. 1984. Belief and the Will. *Journal of Philosophy* 81(5): 235–56.

——. 1989. *Laws and Symmetry*. Oxford: Oxford University Press.

van Lambalgen, Michiel. 1987a. Von Mises' Definition of Random Sequences Reconsidered. *Journal of Symbolic Logic* 52(3): 725–55.

——. 1987b. *Random Sequences*, PhD dissertation, University of Amsterdam. <http://staff.science.uva.nl/~michiell/docs/fFDiss.pdf>.

——. 1996. Randomness and Foundations of Probability: Von Mises' Axiomatisation of Random Sequences. In *Probability, Statistics and Game Theory: Papers in Honor of David Blackwell*, Lecture Notes-Monograph Series, Vol. 30, 347–68. Edited by T. Ferguson, L.S. Shapley, and J.B. MacQueen. Hayward, CA: Institute for Mathematical Statistics.

Venn, John. 1876. *The Logic of Chance*. 2nd edn. London: MacMillan.

Verduin, Kees. 2009. Christiaan Huygens 'Under Construction'. <http://www.leidenuniv.nl/fsw/verduin/stathist/huygproj.htm>.

Ville, Jean. (1939) 2005. A Counterexample to Richard von Mises's Theory of Collectives'. Partial translation of *Étude Critique de la Notion de Collectif* by Glenn Shafer. <http://www.probabilityandfinance.com/misc/ville1939.pdf>.

Vis, M.A. and A.L. Van Gent (eds). 2007. Road Safety Performance Indicators: Country Comparisons. Deliverable D3.7a of the EU FP6 Project SafetyNet. <http://erso.swov.nl/safetynet/fixed/WP3/sn_wp3_d3p7a_spi_country_comparisons.pdf>.

von Mises, Richard. 1938. Quelques Remarques sur les Fondements du Calcul des Probabilités. In *Colloque Consacré a la Théorie des Probabilités*, part 2. *Actualités Scientifiques et Industrielles*, Vol. 737, 57–66. Paris: Hermann & Cie.

——. (1939) 1951 *Positivism*. Translated by J. Bernstein and R.G. Newton. New York: Dover Publications, Inc.

——. (1957) 1981. *Probability, Statistics and Truth*, 2nd English edn. Translated by J. Neyman, D. Scholl, and E. Rabinowitsch from the 3rd 1951 German edition. Edited by Hilda Geiringer. 1st German edn. 1928. Reprint, New York: Dover Publications, Inc.

——. 1964. *Mathematical Theory of Probability and Statistics*. Edited and complemented by Hilda Geiringer. New York: Academic Press.

—— and J.L. Doob. 1941. Discussion of Papers on Probability Theory. *Annals of Mathematical Statistics* 12(2): 215–17.

von Neumann, John and Oskar Morgenstern. 1944. *Theory of Games and Economic Behaviour*. Princeton: Princeton University Press.

Wald, Abraham. 1938. Die Widerspruchsfreiheit des Kollektivbegriffes. In *Colloque Consacré a la Théorie des Probabilités*, part 2. *Actualités Scientifiques et Industrielles*, Vol. 737, 79–99. Paris: Hermann & Cie.

Weyl, Hermann. 1952. *Symmetry*. Princeton: Princeton University Press.

Wheeler, Graham. 1990. *Home Brewing: The CAMRA Guide*. St Albans: Alma Books Ltd.

Williamson, Jon. 2010. *In Defence of Objective Bayesianism*. Oxford: Oxford University Press.

Wright, Georg Hendrik von. 1993. Mach and Musil. In *The Tree of Knowledge and Other Essays*, 53–61. Leiden: Brill.

Zabell, Sandy L. (1989) 2005. The Rule of Succession. In Zabell 2005, 38–73.

——. (1998) 2005. Symmetry and Its Discontents. In Zabell 2005, 3–37.

——. 2005. *Symmetry and Its Discontents: Essays on the History of Inductive Probability*. Cambridge: Cambridge University Press.

Index